AF764275

The Complete Polyethylene Film Extrusion Manual

B.H. Gregory

First published by Plastics Information Direct – a division
of Applied Market Information Ltd – in 2014
Republished by Elsevier

ELSEVIER

AMSTERDAM • BOSTON • HEIDELBERG • LONDON
NEW YORK • OXFORD • PARIS • SAN DIEGO
SAN FRANCISCO • SINGAPORE • SYDNEY • TOKYO

The Complete Polyethylene Film Extrusion Manual

First published by Plastics Information Direct – a division of Applied Market Information Ltd – in 2014
Republished by Elsevier

Elsevier
Radarweg 29, PO Box 211, 1000 AE Amsterdam, Netherlands
The Boulevard, Langford Lane, Kidlington, Oxford OX5 1GB, UK
50 Hampshire St, 5th Floor, Cambridge, MA 02139, USA

Copyright © 2017 Elsevier Ltd

All rights reserved. Except as permitted under current legislation, no part of this publication may be reproduced or distributed in any form or by any means, or be stored in a database or retrieval system without prior written permission of the publisher.

The information provided in this book is based upon the knowledge and opinions of the author. The publishers have made every effort to ensure that the information given in this book is accurate, but no legal responsibility is accepted for any errors or omissions in that information, and no responsibility is accepted for any damages resulting in connection with the use of any information contained herein.

The right of B.H. Gregory to be identified as the author of this work has been asserted by him in accordance with sections 77 and 78 of the Copyright, Designs and Patents Act 1988.

ISBN: 978-1-906479-12-1

Contents

Introduction

This book offers a practical and detailed evaluation of polyethylene film extrusion processes including coextrusion and adhesive lamination. The technology and products are in constant development and it is, therefore, impossible to maintain a continuous update of every current innovation by this rapidly evolving industry. The body of work presented is intended to provide a comprehensive insight to film processors and converters of the various options that are available to produce tailor-made products and how to best exploit the available fabrication technologies.

All the known commercial applications for polyethylene films, coextrusions and laminations are described. Key features of screw and die designs and film treatments, blending and formulations are evaluated.

This study is aimed at a wide readership of technologists, engineers, marketers and students engaged in the development and production of polyolefin films used in packaging, agriculture, construction, consumer, health care and many industrial applications. The aim is to assist in optimizing product performance, evaluating the most cost effective solutions and providing useful information on the key polymers and films commercially available.

Computer modelling

In this study much use is made of computer generated modelling to help in demonstrating and evaluating polymer flows in various parts of the processing equipment. This PC software is designed to simulate non-Newtonian flows in a variety of flow fields including screws, dies and coextrusion processes. The program solves the equations of conservation of mass momentum and energy for polymer melts through process equipment. The finite element method (FEM) is used, which is a rigorous solution of the differential equations. The Galerkin formulation is used to obtain solutions expressed as polynomials within small triangular elements and finally the individual solutions are assembled together to provide the velocity, pressure and temperature throughout the whole field of flow. From the velocity field the streamlines, local shear and stretch rates, stresses and particle residence times are calculated.

Computer programs are also used to predict heat transfer and cooling rates on chill rolls, air gaps and coextrusion dies.

B.H. Gregory
2014

Chapter 1

Polyethylene overview

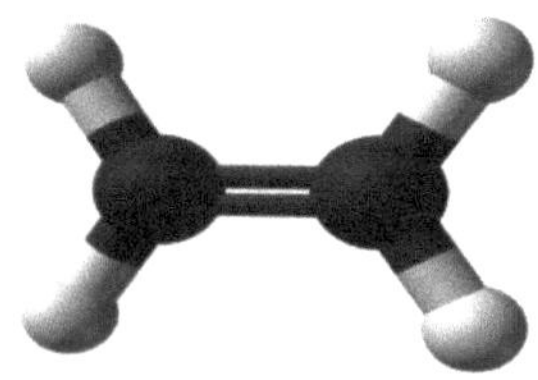

Over the last few decades the polyethylene industry has been in the midst of major restructuring and rationalization. This has resulted in the proliferation of alliances, producer partnerships, joint ventures and acquisitions. Plant productivity is rising while product flexibility is increasing as the process capabilities expand. New catalysts match polymer performance more closely to market needs as new applications are developed. This study reflects these changes and in particular the development of new generation ethylene-based polymers together with innovations in film extrusion, lamination and coextrusion processes.

The annual global polyethylene production capacity is estimated to be 71 million tons (circa 2012).

Five basic commercial polymerization processes are used:

- High pressure free radical initiated autoclave process
- High pressure free radical initiated tubular process
- Gas phase catalytic low pressure process
- Solution catalytic low pressure process
- Slurry (particle form) catalytic low pressure process

The production split between the three key polyethylene categories is as follows:

- LDPE (low density polyethylene) 35%
- HDPE (high density polyethylene) 34%
- LLDPE (linear low density polyethylene) 31%

The high pressure ethylene copolymers are included in the LDPE segment.

1.1 Low density polyethylene (LDPE)

LDPE is defined by a density range of 0.915-0.932 g/cm^3. The polymer embodies a high degree of short and long chain branching, which inhibits the formation of intermolecular attractions. This limits its tensile strength and modulus, but enhances impact resistance. LDPE is polymerized at high temperature and pressure with peroxide initiators that are incorporated into the growing molecular chain and do not leave any residues in the final product. The polymer has excellent flow properties at high shear conditions and is easy to process using all extrusion techniques. Film applications account for 55% of LDPE consumption.

1.2 Linear low density polyethylene (LLDPE)

LLDPE is produced by low pressure processes using multi-site titanium-based Ziegler-Natta (Z-N) catalysts and more recently single-site metallocene catalysts.

LLDPE was defined by the density range of 0.915-0.927 g/cm^3 but the development of metallocene technology has extended the density range below 0.915 g/cm^3. The polymer contains only short chain branches from copolymerization with α-olefins, typically 1-butene, 1-hexene and 1-octene. The polymer has generally superior physical properties to LDPE, but is less sensitive to shear in the melt phase making it more difficult to process. In commercial extrusion processes sharkskin (surface roughness) can occur on film surfaces and its low elongational viscosity can cause bubble instability in the air cooled tubular film process. Film applications account for 40% of LLDPE consumption.

1.3 High density polyethylene (HDPE)

HDPE is defined by a density greater than 0.945 g/cm^3. The polymer contains a low degree of branching and consequently much stronger intermolecular attractions. The polymer has greater rigidity and a higher melting point than the lower density materials.

HDPE can be produced by supported chromium oxide catalysts, Ziegler-Natta catalysts and metallocenes. The very low branching level is achieved by using the appropriate catalyst formulation, reaction conditions and control of comonomer (0.5-3%): usually n-hexene. The slurry loop chromium oxide process produces wide molecular weight distribution polymers, which are easier to process particularly at high molecular weight. Ziegler-Natta catalysts produced polymers with much narrower molecular weight distributions and were at a disadvantage in blow moulding and pipe extrusion applications requiring high melt strength.

This led to the development of dual reactor processes to produce high molecular weight polymers with very wide molecular weight distributions with Z-N catalysts. These polymers can produce very high strength films using the long neck (wine glass) tubular film process (see Section 10.2).

Film applications account for 26% of HDPE consumption.

1.4 Medium density polyethylene (MDPE)

MDPE is defined by a density range of 0.930-0.945 g/cm^3. These can be produced by all catalytic processes at low pressure. MDPE has better stress cracking resistance than HDPE, but a lower tensile strength and stiffness and it has a variety of film applications.

1.5 Metallocene polyethylene (mPE)

The metallocene catalysts widen the range of polyolefin materials by precise control of side chain branching (SCB), which enables the polymerization of a density range from 0.865 to 0.960 g/cm^3, i.e. from elastomeric (POE) polymers to rigid HDPE. Different forms of polypropylenes (PP) have also been synthesized with metallocenes.

The important variable is the control of SCB distribution, usually described as composition distribution (CD). Narrowing the CD distribution results in favourable improvements, such as tensile and impact strength, better opticals, lower melting points and a much reduced level of extractable material.

Metallocene technology is continually evolving as new generation catalysts are developed. Much of the research is focused at improving processability by inducing long chain branching (LCB) and widening molecular weight distributions (MWD) where required, by using series operated multi-reactor trains.

1.6 Ethylene copolymers

Comonomers containing heteroatoms will poison all organometallic (e.g. Ziegler-Natta) catalysts. These copolymers are only produced by the free radical initiated high pressure process and incorporate comonomers such as vinyl acetate, acrylic acid and alkyl acrylates. The presence of an acid or ester functionality enhances specific adhesive properties and miscibility with many organic and inorganic substances. The comonomer reduces crystallinity, which reduces stiffness and melting point. These copolymers are easily processed by all film forming techniques and are used in a wide variety of film, blending, compounding and coating applications. Ethylene-vinyl acetate copolymers (EVA) with 28% vinyl acetate (VA) content are widely used in wax blends and hot melt adhesives.

Chapter 2

Polyethylene molecular structures

Polyethylene (LDPE) invented by ICI in 1933 is produced in very high pressure reactors at high temperature by free radical initiation. Reactor pressure ranges from 1500-2500 bar at temperatures of 250-300 °C. Its molecular structure is complex as it combines both short chain (C_2-C_5) and undetermined long chain branches, which reduce crystallinity to around 50%. The molecular weight distributions are wide, which enabled easy processing in the extrusion and moulding processes of the time. The polymer is inert, free of catalyst residues and heat stable, unlike PVC (polyvinyl chloride), and has attractive physical properties. By the late 1940s organometallic catalysts-based on complexes of titanium halides and aluminium alkyls, developed by Ziegler and Natta, enabled the polymerization of both ethylene and propylene. The ethylene polymers incorporated very low amounts of short chain branching resulting in much higher crystallinity. Concurrently Phillips Petroleum produced similar lightly branched polymers using chromium oxide catalysts. These new polymers were defined as high density polyethylene (HDPE) and had a melting point of 130-133 °C compared to 110-115 °C for the high pressure polymers (LDPE).

During the 1960s, Ziegler-Natta catalysts were developed by DuPont enabling the copolymerization of ethylene with other α-olefins (n-butene and n-octene.) This produced copolymers (LLDPE) similar to high pressure polyethylenes, but with narrower molecular weight distributions (MWD) resulting in a rheology less suited to the existing high volume polyethylene film processes. The metallocene catalysts commercialized in the early 1990s have since unleashed a flood of new polyethylenes focused on the development of products that combine the broad range of attractive physical properties possible with polyethylenes with flow properties that attempt to better match the needs of extrusion processes, such as tubular film production and extrusion coating.

Figure 2.1 visually differentiates the molecular structures of the four basic polyethylene types commercially available. These are:

- LDPE
- LLDPE
- mLLDPE (metallocene-catalyzed)
- HDPE

LDPE has a complex molecular architecture as mentioned earlier that includes both long chain (LCB) and short chain (SCB) branching as shown in **Figure 2.1**. Its broad MWD creates flow properties that are highly sensitive to shear rate in melt processes. This makes LDPE one of the easiest polymers to process at high extruder output. The polyethylenes produced by all the low pressure catalytic processes incorporate only SCB (no measurable LCB) and have relatively narrow molecular weight distributions. The metallocene catalysts have better control of composition distributions (CD) which significantly enhances physical properties and reduces extractables, but MWD is narrowed further.

Figure 2.2 shows typical LDPE molecular weight distribution (MWD) curves. MWD is defined as the ratio of the weight average molecular weight (M_W) to the number average molecular weight (M_n).

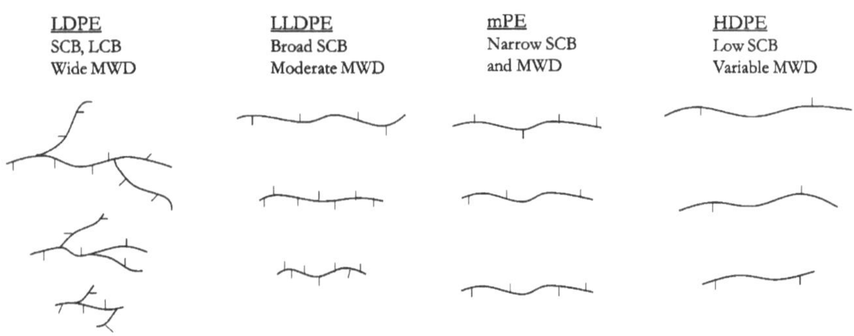

Figure 2.1

Basic molecular structures of polyethylenes

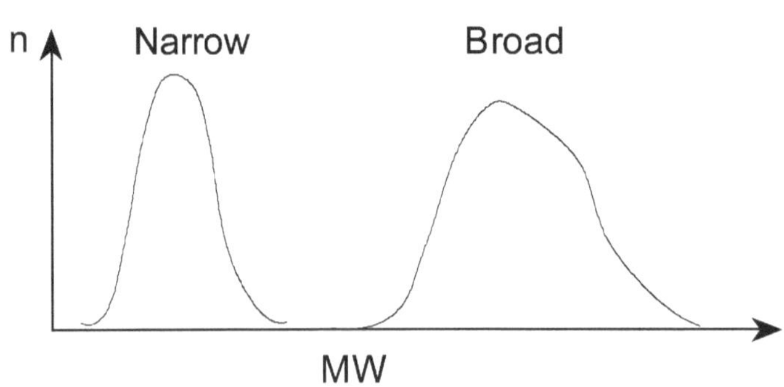

Figure 2.2

Typical molecular weight distribution curves for LDPE

The MWD is called narrow if the polymer is made up of chains close to the average length. If the chain lengths vary widely the polymer is termed as broad or wide in MWD.

The molecular weight distributions for the three basic commercial low density polyethylene processes are shown in **Table 2.1**. The metallocene LLDPEs are the narrowest in MWD.

Table 2.1

Typical molecular weight distributions of low density PEs

Type	Process	Typical MWD
LDPE	High pressure	5-10
LLDPE (Z-N)	Low pressure	3-4
Metallocene LLDPE	Low pressure	~2

Narrow MWD PEs will have better opticals and flex crack resistance and better drawdown than the broad MWD grades. Polymers with broad MWD plus LCB tend to have higher melt elasticity and produce hazier films. The broader the MWD the more shear-sensitive is the polymer in the melt state, which will reduce screw torque and shear stress in extrusion processes. Conversely the narrow MWD polymers are much less sensitive to shear rate leading to high pressures in melt processes.

Chapter 3

Low pressure polyethylenes

LLDPE is produced at low pressure with organo-metallic (Ziegler-Natta) catalysts. The density is controlled by copolymerization with α-olefins. As can be seen from the representation in **Figure 2.1** a wide SCB distribution (CD) is produced resulting in a high level of extractables particularly as the density is lowered.

The higher α-olefins (HAO) used as comonomers with ethylene are listed in **Figure 3.1**. The longest side chain is inserted with n-octene. The octene LLDPEs are made by a solution process. The others can all be copolymerized by either gas phase, solution or slurry technology. The higher the α-olefin content the lower the density.

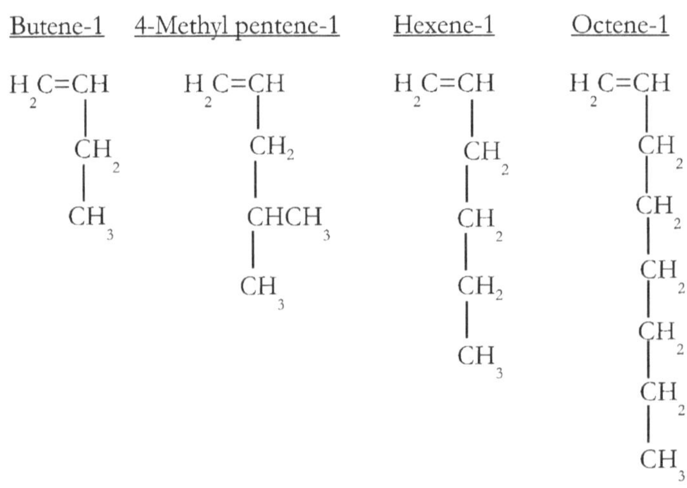

Figure 3.1

Alpha-olefins used in LLDPE production

Table 3.1 compares LLDPE (C_4) to an LDPE of the same melt index and density. The film properties show the LLDPE with superior dart impact, puncture resistance and tensile strength. The lower transverse direction (TD) tear is due to a higher level of machine direction (MD) film orientation during the extrusion process. These property improvements are caused mainly by the narrower MWD.

Table 3.1

Properties of LLDPE vs. LDPE (measured on 37 μm blown film)

Property	Unit	LLDPE (C_4)	LDPE
Melt index	dg/min	1.0	1.0
Density	g/cm^3	0.918	0.918
Dart impact	g	150	90
Puncture	J/mm	68	27
Elmendorf tear MD/TD	g/μm	3.5/9.2	6.3/3.7
Tensile strength MD/TD	MPa	38/31	17/14

In **Figure 3.2** the blown film properties of LLDPEs made with three α-olefin comonomers are compared subjectively. The rule-of-thumb is that the larger the comonomer molecule the better the film properties. This is attributed to the probability of more chain entanglements as the side chain increases in length. The differences between the two higher α-olefins (C_6 and C_8) become increasingly marginal.

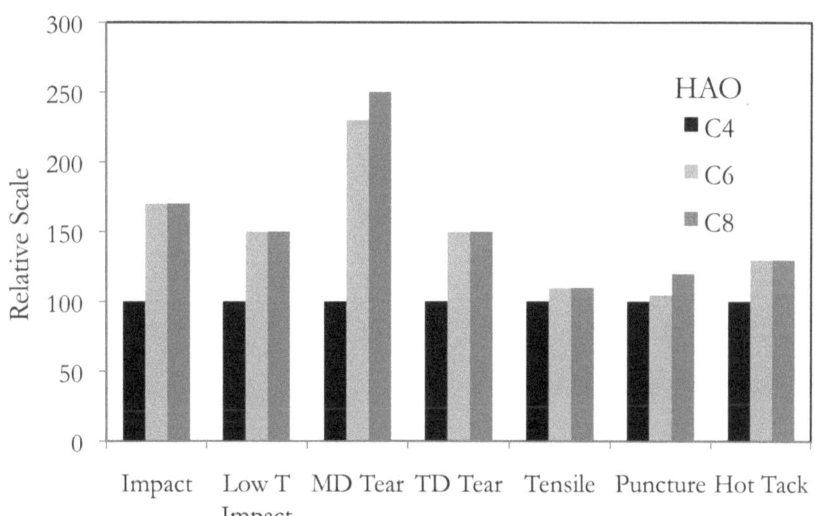

Figure 3.2

Comparative influence of α-olefin comonomer on properties

However, making simple comparisons can be misleading since other structural variables such as morphology, molecular weight and composition distributions (CD) may not have been taken into account. All of these variables will have some effect on both melt flow and solid state properties, which may mask the contribution from the specific comonomer utilized.

As a basic rule the high α-olefin (C_6/C_8) LLDPE grades are superior in ultimate elongation, puncture resistance, tear strength and dart impact to the n-butene grades at equivalent melt index and density. This is reflected in the pricing structures.

The metallocene catalysts, commercialized in the early 1990s, narrowed MWD further and produced very narrow CD (SCB) distributions. The control over the insertion of the α-olefin comonomer significantly improved physical properties. A more homogeneous polymer is produced with reduced low molecular weight (waxy) content.

An example of the improved film impact strengths of 0.920 g/cm^3 density LLDPE produced by metallocene and Z-N catalysts can be as much as:

- mLLDPE = 1000 g
- LLDPE = 300 g

Low extractables (MW) reduces blocking of films. The comparative film blocking forces of two LLDPE grades of 0.920 g/cm^3 density are:

- mLLDPE blocking force = <0.1 g
- LLDPE blocking force = 5 g

This is particularly significant in markets for soft films, food contact and medical applications. On the downside, metallocene polyethylenes suffer from rheology related deficiencies, i.e. melt state properties as listed below:

- High torque and pressure in extruders
- Low melt strength (poor bubble stability)
- Melt fracture/sharkskin

To overcome these shortcomings all the polyethylene producers have focused research to enable the development of mPEs with LCB and wider MWD to improve processability. New generation catalysts have been developed to induce some LCB as well as the use of dual reactor systems to broaden MWD.

Four basic types of metallocene catalyzed linear PEs are currently available commercially as described in **Table 3.2**. Their unique structural homogeneity offers products with significantly enhanced properties. The polyolefin elastomers (POE) are used in TPE (thermoplastic elastomer) applications.

Table 3.2

Basic categories of metallocene PE

Type	Density (g/cm^3)	Description
mMDPE	0.930-0.940	Medium density mPE
mLLDPE	> 0.915	Competitive with LLDPE (7-10% α-olefin)
Very low density	0.895-0.915	α-olefin <20%, competes with EVAs etc.
Polyolefin elastomer	0.865-0.895	α-olefin > 20%, TPE candidate

Chapter 4

Stress in shear and elongational flow

LDPE is unique amongst the polyethylene family in that its melt viscosity in elongational flow is strain hardening. This means that as the melt is stretched its elongational viscosity will increase with increasing strain rate. LLDPE and HDPE and polymers such as polypropylene, polyamides and polyesters are in contrast described as tension thinning. Tension thinning occurs when the elongational viscosity drops with increasing extension rate and the extrudate is readily drawn to very thin sections. This enables HDPE and PP to be fibre forming, whereas LDPE cannot be drawn into highly oriented filaments or films. At fairly low strain levels, the elongational viscosity of LDPE will increase rapidly leading to cohesive failure, i.e. melt break. These two types of extensional flow behaviour are shown schematically in **Figure 4.1**. The ringed section is within the strain rate range of most extrusion processes.

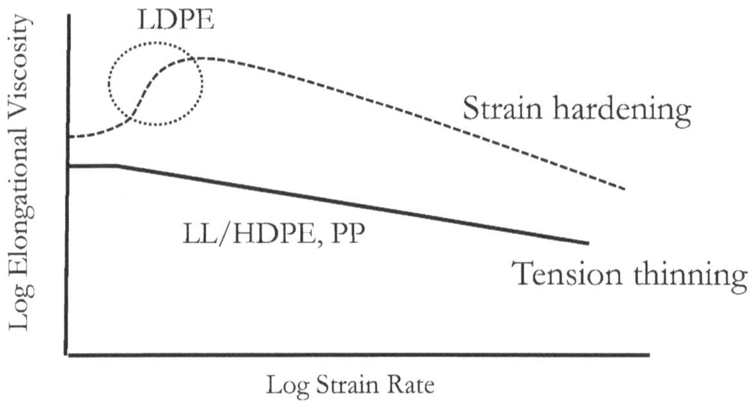

Figure 4.1

Strain hardening and tension thinning

For the production of blown film, it is important to maintain a high elongational viscosity in the gap between the die exit and the frost line of the bubble. This is essential for bubble stability. In addition, the high shear sensitivity of LDPE reduces the possibility of melt fracture at high extruder outputs. These rheological characteristics makes LDPE the ideal polymer for the blown film process. In contrast the linear polymers: LLDPE and PP, are best suited to the cast film process where low extensional viscosity is preferred. HDPE is the exception because of the ability to synthesize very wide MWD, particularly with cascade reactor processes, which simultaneously enables high melt strength with low shear viscosity. The development of HDPE with bimodal distributions and high molecular weight (HMW) has opened up large volume film applications in consumer markets. Bimodal distribution is discussed further in Section 5.1.

This critical difference between LLDPE, including metallocenes, and LDPE is in their relative ease of extrusion. This is demonstrated in **Figure 4.2** where the apparent shear rate dependence of melt viscosity of ~1.0 dg/min. melt index LLDPE, mLLDPE and LDPE are compared at 230 °C. The mLLDPE has the highest shear viscosity.

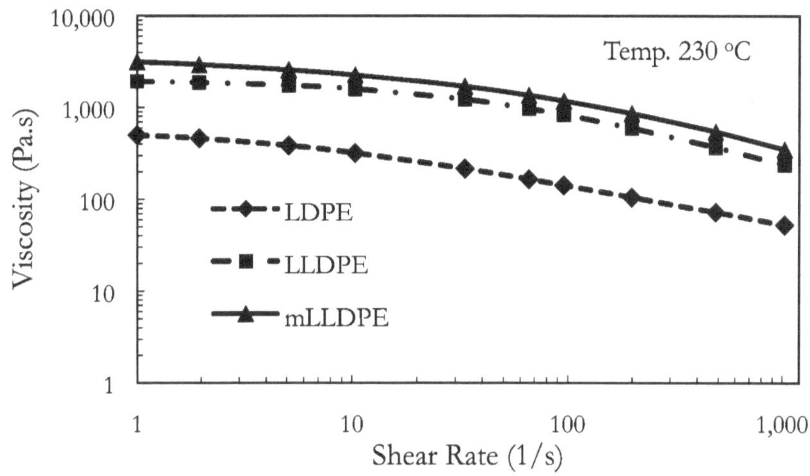

Figure 4.2

Shear viscosity of LDPE and LLDPE

Figure 4.3 compares pressures in a typical 114 mm (4½″) extruder. There is an approximate three-fold difference in peak pressure between the LDPE and LLDPE. Extruder design modifications are necessary to avoid excessive loads and pressure in the extruder and die.

A simple calculation comparing the flow through an annular channel typical of the lips of a blown film die illustrates this crucial difference between LDPE and LLDPE. The melt index is ~1 dg/min. for both resins. The annular die dimensions and inputs are as shown in **Table 4.1(a)**.

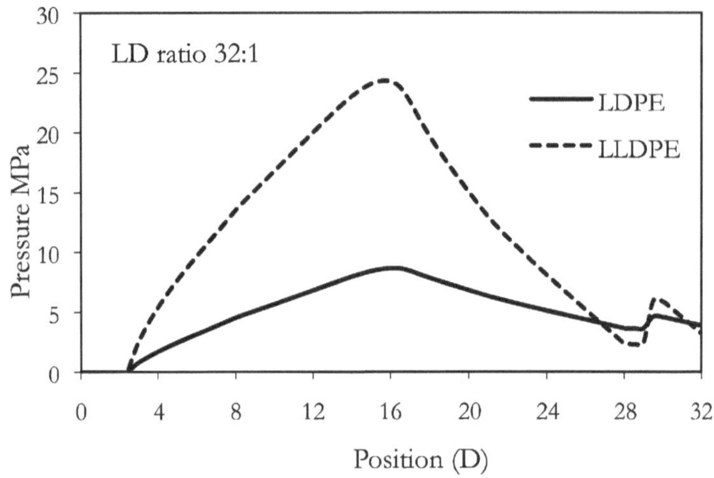

Figure 4.3

Comparative screw pressure profiles

Table 4.1(a)
Die dimensions and inputs

OD diameter	303 mm
ID diameter	300 mm
Land	10 mm
Die gap	1.5 mm
Flow rate	200 kg/hr
Temperature	220 °C

Table 4.1(b) summarizes the flow behaviour of the two polymers through this simple geometry. The calculation is based on comparative capillary viscosity (Carreau model) data.

Table 4.1(b)
Comparative flow of LDPE and LLDPE in annular dies

Flow	LDPE (MI 0.8 dg/min)	LLDPE (MI 1.0 dg/min)
Mass flow rate (kg/hr)	200	
Melt temp. (°C)	220	
ΔP (MPa)	1.0	3.2
Shear stress at walls (kPa)	75.3	238

The pressure (ΔP) is threefold higher for the LLDPE, which is reflected in shear stresses at the walls of 238 kPa versus 75 kPa for the LDPE. This excessive shear stress will result in sharkskin. Widening the die gap is the easiest remedy to eliminate surface melt fracture. However, this solution will result in unbalanced orientation in the film and poorer thickness distribution from the die. Processors have developed compromise methods, such as blending with LDPE and/or the use of polymer processing aids (PPA) and localized heating of the die lips amongst others to reduce shear stress at the lips.

The challenge for the polyethylene producers is to develop polymers that alleviate these limitations without the need for blending, PPA addition etc.

Chapter 5

High density polyethylene films

The development of the long neck blown film extrusion process with high blow-up ratio has made it possible to produce high strength low gauge HDPE films, which compete with LDPE, LLDPE and paper in commodity bag, sack and bin liner markets. For the process to work effectively, HDPEs of high melt strength and toughness are required. This is achieved by the polymerization of a high molecular weight (HMW) fraction for strength, combined with very broad MWDs for processability. Splittiness in thin films is avoided by balancing the MD/TD orientation during the tubular extrusion process.

HDPE films have the following distinctive advantages:

- High strength to thickness ratio, i.e. reduced cost and lower carbon footprint
- Continual improvements in extrusion equipment to allow wider web production, better thickness control and higher output, e.g. grooved barrel feed sections
- Improved grades: particularly bimodal HMW-HDPE
- Resource reduction, i.e. downgauging
- Non-blocking: allows easy openability of paper-like HDPE bags

Other major applications for HMW-HDPE are in blow moulding of industrial containers and extrusion of pressure pipes.

For the best combination of processability and strength, the Z-N catalyst multiple stirred tank slurry processes have been very successful in the manufacture of HMW-HDPE film grade resins. In slurry processes the polymer separates from the carrier as fine particles, which facilitates the production of HMW polymer. However, Z-N catalysts produce narrow MWD HDPEs. To overcome this limitation two-stage series reactors are used. These synthesise very wide MWD (>15:1) polymers. In the first reactor the low MW fraction is produced and the active particles are transferred into the second reactor to polymerize the high molecular weight component. A broad range of MWDs can be produced with this slurry technology in both unimodal and bimodal distributions.

The alternate supported chromium (Cr) oxide catalyst slurry process (Phillips) produces MWD of approximately 12:1 in single stage slurry loop reactors.

These HDPEs are widely used in blow moulding, but less popular in blown film applications. However, new generation grades using this process have improved significantly in recent years and are claimed by the manufacturers to be competitive in many film applications.

The latest technologies are capable of being adapted to "swing" plant operation to produce both LLDPE and HDPE in the same reactor. Chevron Phillips claims that their chromium oxide-based slurry loop system can convert from HDPE to LLDPE within a few hours, whereas gas phase processes may require a couple of days to be on prime quality.

HDPE resins for blown film applications are divided into the following molecular weight categories. Melting points range from 130-133 °C.

HDPE general purpose	<180,000
Medium molecular weight (MMW)	180,000-280,000
High molecular weight (HMW)	300,000-400,000

In film markets, MMW-HDPE resins are used in snack food packaging (often in a coextruded structure) and a variety of merchandise and small bag applications. HMW-HDPEs are used for their high end use performance properties of high impact and tensile strength and low creep in applications such as shopping bags, trash can liners and garbage bags. The very high strength of HMW-HDPE allows for significant downgauging, cost savings and reduced carbon footprint.

Typical molecular weight distributions for HDPE resins used in blown film processes are as follows.

Medium	12-20:1
Broad/bimodal	20-40:1

These MWD values are appreciably higher than those normally associated with the lower density polyethylenes. The basic characteristics of HDPE grades by key application used in blown film applications are shown in **Table 5.1**. HMW-HDPE resins can accept high concentrations of recyclates.

A total absence of gels and other hard particles in the polymer is a critical requirement for thin film extrusion. All HDPE products intended for blown film extrusion must be clean and gel-free to avoid holes and breaks in the bubble.

The attraction of HDPE films is partly due to their crisp paper-like feel and opacity, which can be augmented by fillers ($CaCO_3$) and pigments (TiO_2). Filler addition is around 5% added from high concentration masterbatches. These films have excellent openability and they handle well on high speed converting and bag making machines. High moisture and grease resistance are other valuable attributes. The ability to significantly downgauge has been the primary driving force in the rapid growth in demand for HDPE films in bag and liner applications.

Table 5.1
HDPE used in blown film extrusion

Applications	Melt flow rate (kg) @ 230 °C				MWD (M_w/M_n)	Density (g/cm³)
	2.16 kg	5 kg	10 kg	21.6 kg		
Medium density	0.17				Medium	0.938
High impact		0.2		6	Broad	0.945
Standard HMW		0.3		5-9	Broad	0.953
General purpose	0.3	1.4			Med./broad	0.955
Quality grade		0.3		9	Broad	0.955
General purpose	0.3	1.5			Medium	0.959
Moisture barrier	0.45		5	20	Medium	0.962

5.1 Bimodal HMW-HDPE

The development of grooved (feed bushing) barrel extruders broadened the scope for the use of high molecular weight HDPE in the blown film process. However the processable polymers were still restricted to grades with relatively high melt flow rates thus limiting the strength related properties. The density was usually below 0.952 g/cm³ to maximize impact strength at low gauge. Typical unimodal broad MWD resins produced films with acceptable toughness at a thickness of 25 μm and above, but lacked sufficient strength at lower gauges. Using advanced Z-N catalysts bimodal HDPE grades were developed as a means of maximizing strength, stiffness and drawdown whilst allowing downgauging to 12 μm and below. The intent was to open up large volume opportunities for HDPE films to replace LL/LDPE and kraft paper in bag and sack markets.

In **Figure 5.1** the molecular weight distribution of a typical wide MWD HDPE is compared to an idealized bimodal distribution. The bimodal polymer has two distinct peaks controlled by the catalyst used in the polymerization. This is achieved with proprietary Z-N catalyst formulations and series reactors. The peaks at the high and low end of the distribution maximize strength and flow characteristics respectively. The MWD and SCB is controlled by dual polymerization reactors using one catalyst. The concentration and location of the α-olefin comonomer incorporation is critical. Densities of bimodal grades range from 0.945 to 0.956 g/cm^3. These densities will result in stiff paper-like films of very high strength.

Recent advances in metallocene catalysts have given the capability to produce bimodal distributions in single stage processes, which should be less costly than the dual reactor processes using conventional catalysts. These developments should provide new opportunities for polyethylene processors.

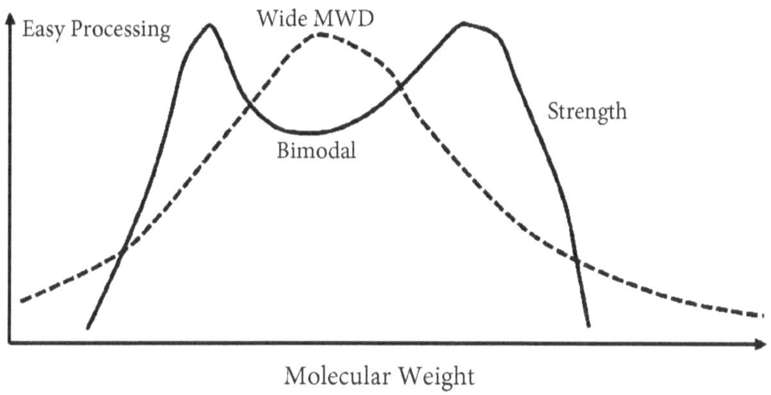

Figure 5.1

Bimodal HMW-HDPE

Superior physical properties of bimodal resins are a result of an increase in the number of tie-molecules of the same melt index and density. It is important to insert the α-olefin comonomer in the high MW fractions to allow the maximum formation of tie-molecules. The formation of tie-molecules is visualized in **Figure 5.2**. The series reactor slurry process increases the number of tie-molecules shown in **Figure 5.2** thus enhancing strength properties required in low gauge applications such as T-shirt grocery bags.

Bimodal HDPE is less sensitive to MD orientation and as a consequence produces more balanced films at any given blow-up ratio. This partly explains the very high impact strengths achieved with thin films. Some basic characteristics of unimodal and bimodal HMW-HDPE grades prepared with two catalysts are summarized in **Table 5.2**.

The uniqueness of the bimodal distribution is best demonstrated by comparing shear viscosities and their impact on shear stress at die walls. **Figure 5.3** plots the flow curves of a broad MWD HMW-HDPE, with a HMW bimodal HDPE. The HDPE has a melt index of 0.1 dg/min. and is produced with a chromium catalyst in a slurry loop reactor. An LLDPE (C_8) produced with Z-N catalyst in a solution process is also included. This is a general purpose grade of melt index 1.0 dg/min.

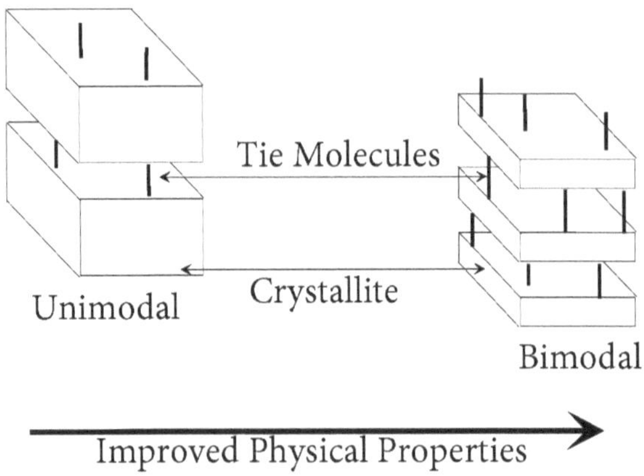

Figure 5.2

Tie-molecules in HDPE

Table 5.2
Characterization of HDPE

HMW-HDPE	Unimodal	Bimodal
Catalyst type	Cr	Ti (Z-N)
M_w/M_n	10-20	20-40
M_w	<250,000	>300,000
% Crystallinity	60-75	60-75
% CHMWS[1]	~15	>20
% CLMWS[2]	~1	<1

1: conc. HMW species >500,000.
2: conc. LMW species <1000.

The flow curves show that in the typical film processing range of 100-500 s^{-1} shear rate the bimodal HDPE has a lower viscosity than either the conventional HMW-HDPE or the LLDPE. This is very significant. This data was used to estimate the shear stress developed in a die of 250 mm diameter with a land length at the lips of 5 mm. Mass flow rate was 180 kg/hr at 230 °C. **Figure 5.4** summarizes the results at three die gaps.

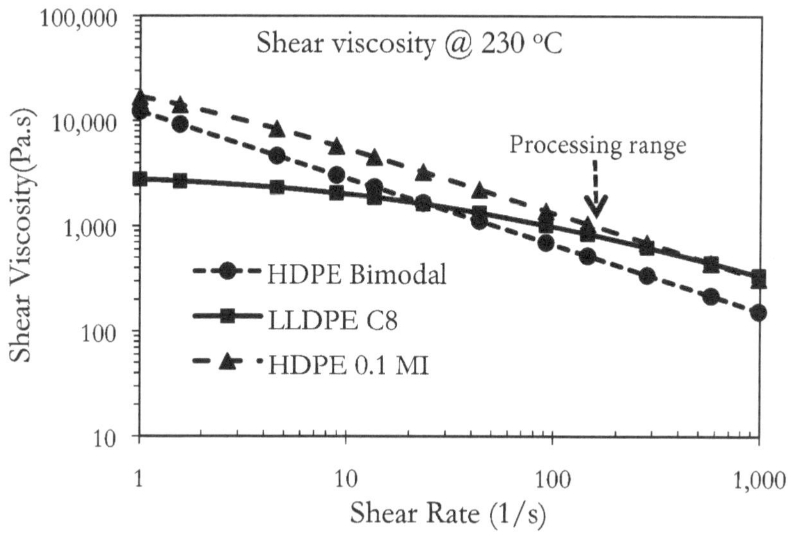

Figure 5.3

Comparative shear viscosity data

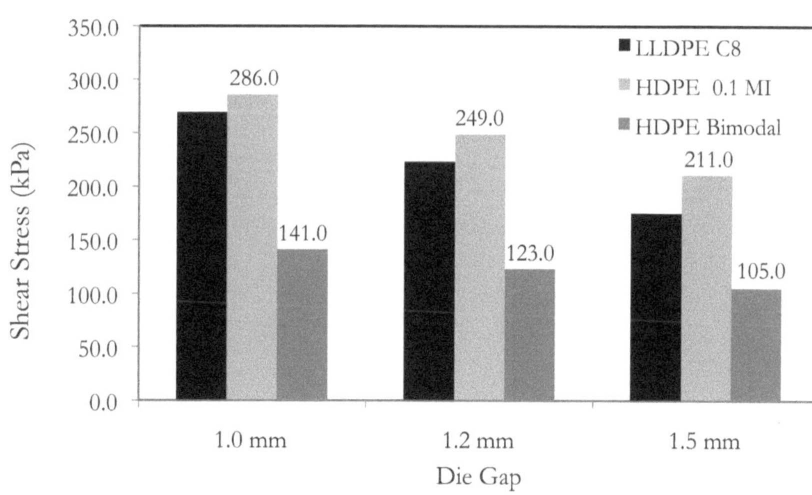

Figure 5.4

Shear stress in 250 mm tubular die

Assuming a sharkskin threshold of 140 kPa the bimodal HDPE can probably be run with a die gap of 1.0-1.2 mm. The other two resins will require die gaps well above 1.5 mm probably in the range of 2.0-2.5 mm. These simulations are very striking particularly for the LLDPE, which has by far the highest melt index. Both HDPEs have much wider distributions than the LLDPE, which clearly affects the flow properties. For the processor narrow die gaps are preferred to control thickness distribution and minimize excessive drawdown in the MD.

Film property comparisons of unimodal and bimodal HMW-HDPE resins are shown in **Table 5.3**. The improved impact strength and tear resistance of the bimodal polymer are clearly illustrated. The availability of bimodal HMW-HDPE has made it possible to downgauge grocery bags from 20-25 μm to 12-15 μm. This has a profound effect on costs as well as a reduction in carbon footprint.

Table 5.3

Comparison of bimodal and unimodal HMW-HDPE (blow-up ratio 4:1)

Property	Unit	Bimodal	Unimodal
MFR_5	dg/min	0.31	0.4
Density	g/cm^3	0.944	0.948
Dart impact strength	g/12 μm	320	140
Dart impact strength	g/25 μm	425	225
Tear MD/TD	g/25 μm	32/98	19/46

Bimodal HDPE has excellent compatibility with other polyolefins. This is an advantage when blending with recyclates. There have been many claims made regarding the superior results obtained by blending up to 50% recyclate with bimodal HMW-HDPE compared to conventional resins.

Chapter 6

High pressure ethylene copolymers

The high pressure LDPE process can tolerate comonomers containing heteroatoms. This allows ethylene copolymers with ester or acid functionality or both to be produced. These copolymers like their homopolymer analogues will contain both long and short chain branching in addition to the comonomer side chain. In contrast transition metal catalysts, such as Ziegler-Natta, used in the low pressure processes are poisoned by the presence of oxygen in the comonomers. The metallocene catalysts also suffer from the same limitation.

The best known ethylene copolymers are listed below with abbreviations and percentage comonomer:

Ethylene-vinyl acetate	EVA	2-40%
Ethylene-acrylic acid	EAA	2-9%
Ethylene-methacrylic acid	EMAA	2-9%
Ethylene-methyl acrylate	EMA	5-25%
Ethylene-ethyl acrylate	EEA	5-25%
Ethylene-n-butyl acrylate	EnBA	5-25%

The comonomer concentration can be varied from 1-40% by weight. The basic structure is shown schematically in **Figure 6.1**.

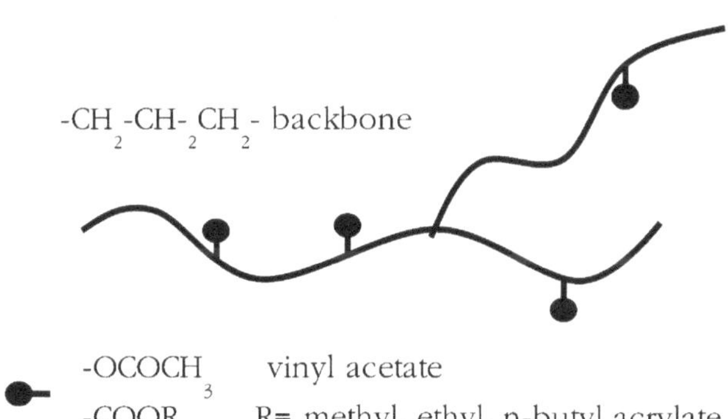

Figure 6.1

Schematic of comonomer distribution on polymer chain

Comonomer addition has a number of effects on properties. Reduced crystallinity improves the following:

- Resistance to cold flexing
- Impact resistance, particularly at low temperatures
- Flexibility
- Stress crack resistance
- Transparency
- Sealability (lower seal temperature)

The following property values are decreased:

- Hardness
- Stiffness
- Heat distortion resistance
- Chemical and solvent resistance
- Barrier to gases

The effect of increased polarity is to improve:

- Filler acceptance
- Adhesion
- Crosslinking (due to more tertiary C atoms)

The drop in crystallinity with VA addition is shown in **Figure 6.2**. Similar changes in crystallinity occur with the acrylates and acid copolymers.

The polar groups, particularly the acetoxy group in EVA, significantly improve compatibility with inorganic materials such as chalk, carbon black, titanium oxide, silica etc.

However there are differences in heat stability between the EVAs and the alkyl acrylates. **Figure 6.3** illustrates the relatively weak oxygen link between the ethylene chain and the acetoxy group. Breakdown occurs at the oxygen atom in the structure to form acetic acid and unsaturation in the residual chain, which can then crosslink. In contrast the acrylate in the EMA is linked to the chain through the much stronger -C-C- bond and consequently this class of copolymer can tolerate higher temperatures.

Figure 6.4 shows the effect of repetitive extrusions on the degradation (melt index change) of EVA versus the more stable EMA. The EVA (2 dg/min MI) incorporates

18% VA and the EMA (6 dg/min MI) contains 21% MA. In **Figure 6.4**, the "new" melt index is divided by the original value. The EVA is dropping in melt index as it crosslinks with each extrusion pass. The EMA shows a slight increase in melt index after several passes at 300 °C extrusion temperature. EVAs should not be extruded at temperatures above 240 °C.

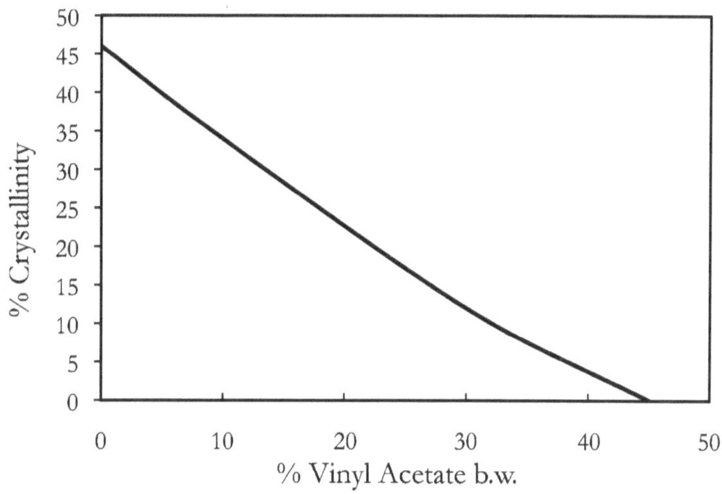

Figure 6.2

Effect of comonomer on crystallinity

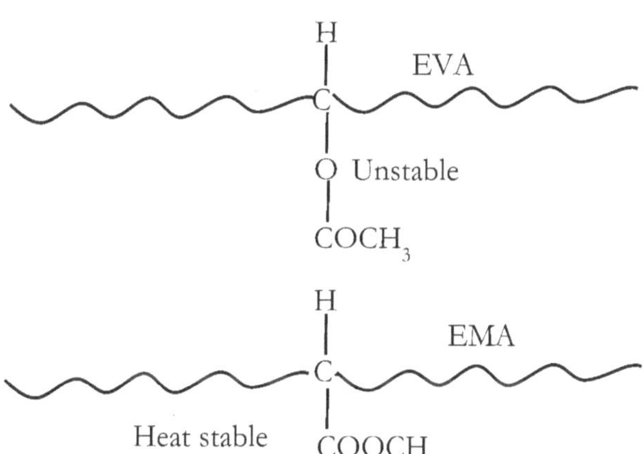

Figure 6.3

Comparison of vinyl acetate and acrylate structures

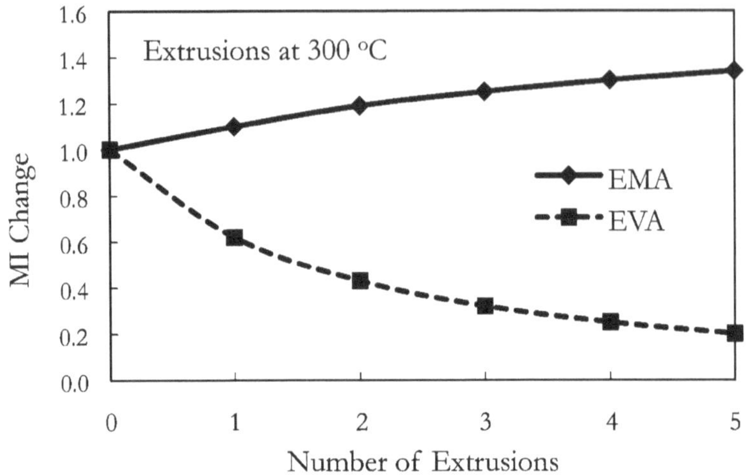

Figure 6.4

Effect of repeated extrusion on degradation (melt index change) of EVA and EMA

EVA films are used in a variety of film applications that include greenhouse shrouds, adhesive lamination, sealable lidding and stretch wrap films where improved heat-sealability, high impact strength and low temperature resistance are required. Optical properties improve with increasing VA levels. The EVAs have better resistance to UV radiation than the homopolymers.

The acrylate copolymers are commercially produced in both autoclave and tubular reactors. Improved technologies enable the production of fractional melt index EMAs with up to 25% comonomer. These are very soft materials that retain some 15% crystallinity with 25% MA incorporation. This copolymer has a melting point of 90 °C, which is claimed to be higher than low density mPEs (metallocene) at the same stiffness. Another advantage is low blocking of the film. The process is claimed to result in a controlled distribution of the MA in the polymer chains and the small size of the MA molecule allows for very close molecular packing. Excellent adhesion to other polyolefins is claimed.

EMA films are also used to produce disposable gloves for sanitary uses and in extrusion coating on polypropylene fabrics (raffia) and carpet backing. EMA and EnBA copolymers of fractional melt index (<1.0 dg/min) can be blended with other polyethylenes to either increase flexibility or improve film impact strength.

6.1 Ethylene acid copolymers (EAA, EMAA)

This class of highly polar copolymers are also produced by the high pressure LDPE process using either acrylic acid (AA) or methacrylic acid (MAA) as comonomer. These copolymers contain hydrogen bonds in their matrix, which enhances their physical properties. The carboxylic acid incorporation does not normally exceed 9% by weight for products intended for film forming processes. The effect of acrylic acid incorporation on stiffness (E-modulus) is shown in **Figure 6.5** and compared with vinyl acetate. The data shows that the EAA copolymer is stiffer than the EVA over the same level of comonomer addition. The presence of hydrogen bonds in the carboxylic acid copolymer counteracts to some degree the softening effect of the comonomer.

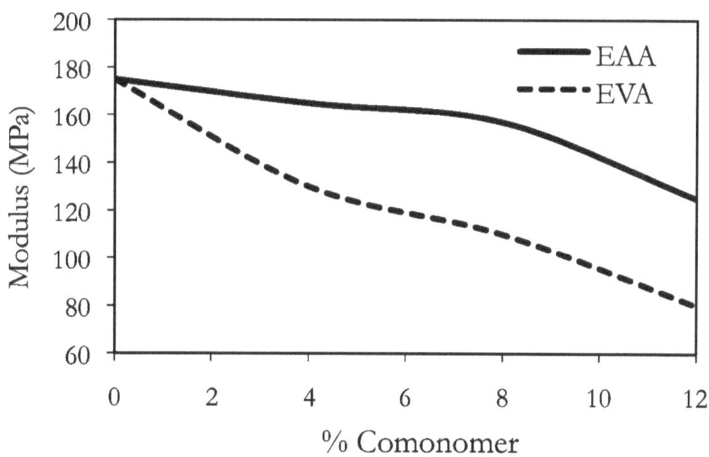

Figure 6.5

Effect of comonomer on modulus

EAA and EMAA copolymers have excellent adhesion to metal surfaces and are very often used in laminates with aluminium foil. The higher the acid content, the higher the polarity and the lower the crystallinity. In addition the degree of hydrogen bonding is increased. The hydrogen bonds form a reinforcing network as mentioned above, which enhances the mechanical properties and melt strength of the polymer and improves hot tack in the heat-sealing process.

The key properties of EAA and EMAA are as follows:

- Excellent adhesion to metals and paper
- Good optics

- Toughness
- Good oil and grease resistance
- Excellent hot tack
- Sealability through contaminants
- High stress crack resistance
- High impact resistance
- High melt strength (low neck-in/good bubble stability)

EAAs are the preferred materials for coating on aluminium foil bearing laminates. Acid levels of 9-12% will have superior adhesion to lower contents of 3-6%. However, the adhesion to polyethylene drops at high acid levels in coextrusions of these two materials.

The preferred usage of EAA is in coextrusion coating with LDPE to maximize adhesion and product resistance. Aseptic packaging of milk and juice products is a large market segment.

Other applications are for shampoo, toothpaste, fresheners, condiments and various pouches and sachets. Their excellent heat-seal and hot tack mean they can be used as seal layers for snack food structures with metallized films or smooth papers.

6.2 Ionomers

The ability to copolymerize ethylene with carboxylic acids led to the invention of ionomers by DuPont. Ionomers are produced from ethylene carboxylic acid copolymers (EAA or EMAA) partially neutralized with reagents containing a metal cation (Zn^{++}, Na^{+}). Reversible ionic crosslinks are introduced in the polymer matrix, which increases toughness and stiffness, improves thermoformability and transparency and develops excellent hot tack in heat-sealing. Ionic crosslinking induces the benefits of crosslinking without the disadvantage of thermosetting the polymer. The ionic matrix weakens as temperature is raised thus allowing the ionomer to flow like a thermoplastic. On cooling the rigid ionic structure is restored.

The acid copolymer (**Figure 6.6**) is neutralized in a melt process with reagents bearing either zinc or sodium cations. These powerful cations form ionic crosslinks, which develop unique properties. Ionomers are moisture-sensitive, particularly the sodium grades, and must be packed in moistureproof containers. Films extruded from these copolymers have excellent impact and penetration resistance and outstanding heat-seal characteristics.

Being a copolymer melting point and stiffness are reduced as crystallinity is depressed by the presence of the carboxylic acid comonomer. However, ionomers are unique materials since they contain some hydrogen bonds formed by residual non-neutralized acid moieties and crucially they embody strong ionic links that result from the presence of powerful cations. In **Figure 6.7**, the increase in stiffness (E-modulus) and impact strength as a function of neutralization (formation of ionic links) is shown. The more cation is incorporated the tougher and stiffer the film becomes.

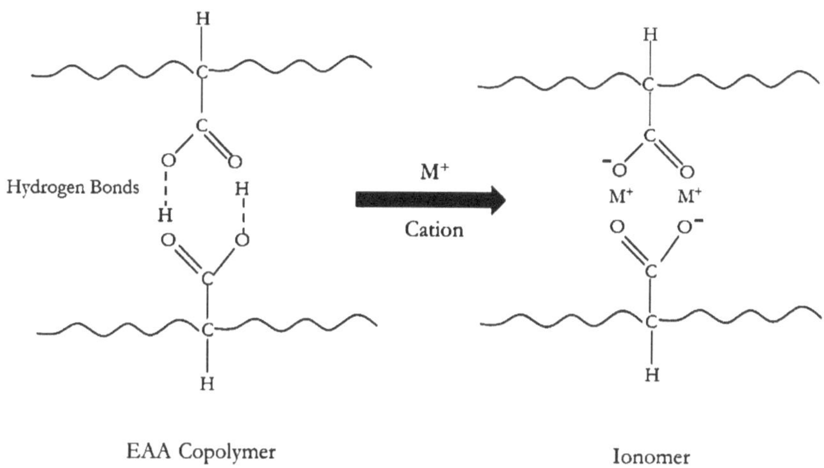

Figure 6.6

Ionomer formation from EAA

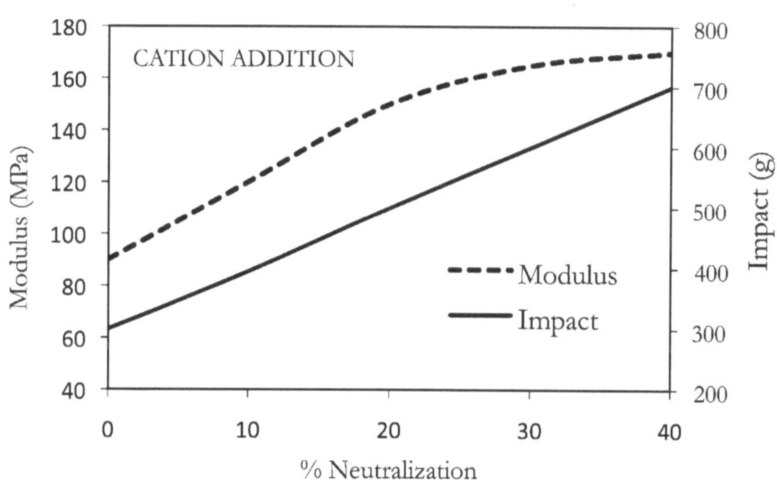

Figure 6.7

Effect of neutralization by cation addition on mechanical properties

The net effect is a material that embodies the following significant property improvements:

- Very high melt strength
- Outstanding penetration resistance
- High-moderate stiffness
- Good oil/grease resistance
- Wide sealing window
- High clarity
- Low sealing temperature
- Seals through contaminants
- Excellent thermoformability

Ionomers are used in blown and cast film processes and the zinc version is used in extrusion coating where it has excellent adhesion to aluminium foil. The zinc ionomer can also be coextruded with polyamide without the need of a tie-resin. In the blown film process their very high melt strength ensures excellent bubble stability. Since the precursor EAA or EMAA are produced by the LDPE high pressure process these resins have relatively wide MWD plus some long chain branching, which results in a shear viscosity similar to LDPE. Therefore, the melt fracture related problems common to linear polyethylenes at high extruder throughputs do not occur with ionomers. The high melt strength results in films with excellent deep draw thermoforming properties.

An interesting feature of ionomers is that they combine very high impact strength with high stiffness. Ionomers are comparable to medium density polyethylene (MDPE) in their rigidity, but much tougher.

The oil resistance of a sodium ionomer is compared with HDPE and EAA in **Table 6.1**. HDPE although having by far the highest crystallinity has the poorer oil resistance. The presence of sodium cation in the ionomer enhances its oil resistance compared to the EAA.

In **Table 6.2** the relative thermoformability, as measured by melt strength, comparing an ionomer to other PE resins is shown. The significantly higher melt strength of the ionomer confirms its excellent performance in deep draw applications. LLDPE, with its narrower MWD and lack of LCB has the lowest melt strength.

Table 6.1

Comparative oil resistance (25 μm films)

Polymer	Oil resistance (hours)	
	Olive oil	Peanut oil
Ionomer (Na^+)	200	180
EAA (9% AA)	60	75
HDPE	40	25

Table 6.2

Relative thermoformability indicated by greatest melt strength

Polymer	Melt strength 190°C
LDPE	20
LLDPE (Z-N)	6
EVA (9% VA)	20
Ionomer	41

Laminates or coextrusions of ionomer with polyamide yield films with excellent deep draw thermoformability and heat-sealing characteristics. These structures will also exhibit outstanding puncture resistance. Ionomers can also be used as impact modifiers for polyamides.

The sodium ionomers have the best oil resistance and clarity. The zinc ionomers have better adhesion to other materials such as PE and polyamide and are less moisture-sensitive. Grades formulated with additives for slip and antiblock are available.

Finally, it is claimed that ionomers can be blended with EVOH (high barrier) to reduce costs. The effects on the gas barrier of the latter are not clear.

Chapter 7

Chemically modified polyolefins

Methods to chemically modify polyolefins by post-polymerization grafting of a reactive functionality on the main chain backbone have evolved rapidly in recent years. The grafting reactions are carried out either in solution or more typically in the melt at elevated temperature. An unsaturated monomer, usually a carboxylic acid, is grafted to the polyolefin by initiating free radicals in the system. In the presence of peroxide initiators, and in some instances only heat, free radicals are formed by chain scission and cleavage of double bonds. The unsaturated monomer then grafts itself onto the main chain backbone. Maleic anhydride is the most widely used reagent although himic anhydride and acrylic acid have also been evaluated in these reactions. Approximately ½% by weight of grafted monomer is sufficient to significantly improve the adhesion and compatibility of the polyolefin to polar materials. The grafted functional group reduces interfacial tension, i.e. acts as a surfactant, which enhances the interaction between immiscible materials. The simplified grafting reaction is shown in **Figure 7.1**. The initiator, organic peroxide, forms a free radical by heating and activates the polymer by hydrogen abstraction. The skill in this technology is to avoid crosslinking or drastic chain scission of the polymer. A monomer which does not easily homopolymerize, such as maleic anhydride, is preferred over acrylic acid which can form polyacrylic acid.

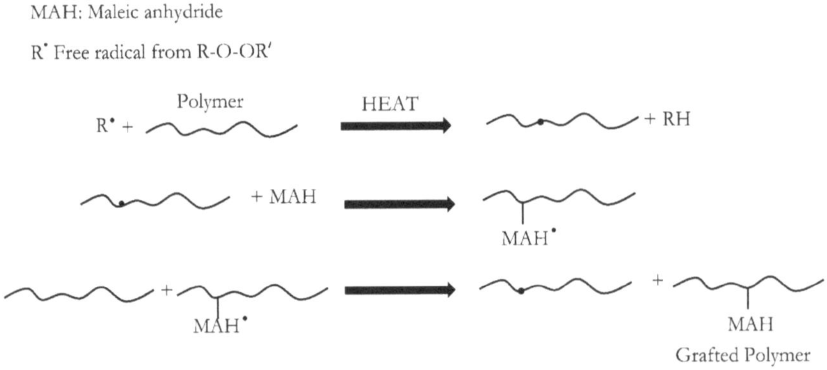

Figure 7.1

Grafting of functional groups on a polyolefin backbone

The grafted polymer is visualized in **Figure 7.2**.

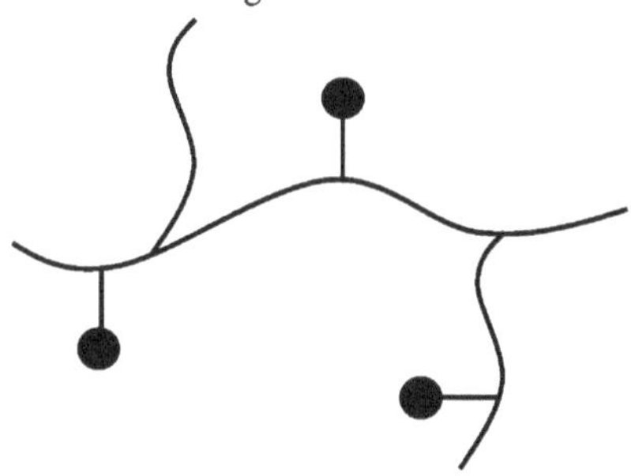

Figure 7.2

Grafted polymer

There are profound differences between PEs and polypropylene (PP) regarding the grafting process. PE tends to crosslink whereas PP undergoes chain scission in the presence of heat and organic peroxides. Controlling these reactions requires specialized skills in the design and operation of the melt extrusion process. The MAH should be randomly distributed on the polymer chains for maximum effect. It is important to protect MAH from moisture as the anhydride is susceptible to hydrolysis.

A schematic representation of the melt extrusion grafting process is shown in **Figure 7.3**. It is recommended to introduce an inert gas (N_2) in the system to avoid oxidation. Twin screw extruders are usually recommended for chemical modification in the melt state. There is, however, some controversy concerning the merits of co-rotating versus counterrotating screws. Both systems are claimed to have been used successfully. Avoiding the formation of gels, i.e. crosslinking is a critical part of the process.

The base polymer can be any polyolefin including ethylene-propylene elastomers (EPDM). An important advantage of this form of modification is that the backbone polymer does not lose any of its key properties. Copolymerization, as discussed earlier, will disrupt crystallinity leading to a reduction in melting point, stiffness and chemical resistance. **Figure 7.4** illustrates the adhesion process at the interface of a coextrusion of MAH-grafted polyolefin to polyamide or ethylene vinyl alcohol (EVOH) polymer and a polyolefin layer. The MAH forms attractive bonds with hydroxyl groups in EVOH or amine groups in a polyamide, and the backbone polymer chain intermixes with the polyolefin layer by diffusion.

LLDPE and HDPE are the easiest of the polyolefins to graft with maleic anhydride and polypropylene the most difficult. The presence of the α-olefin side chain in

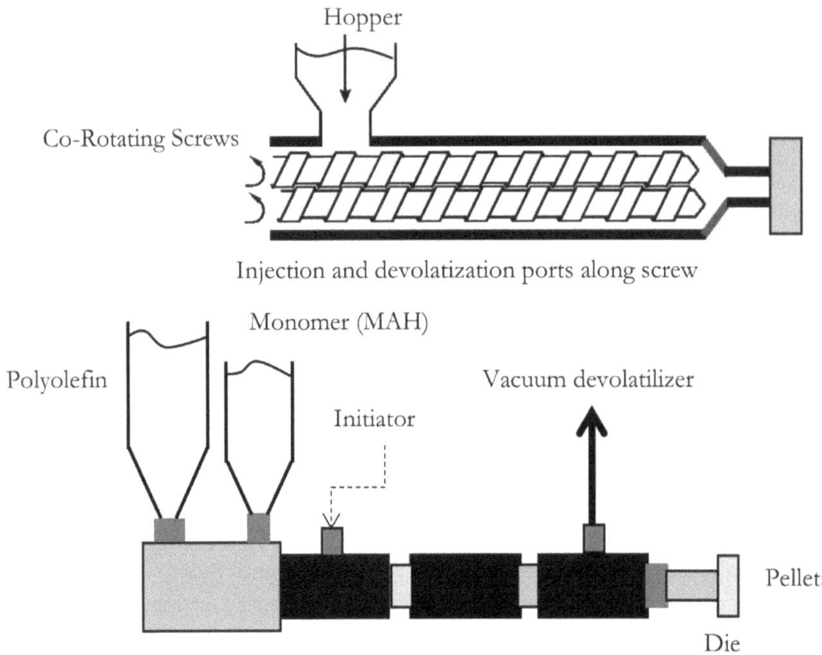

Figure 7.3

Schematic of extruder grafting process using a twin screw extruder

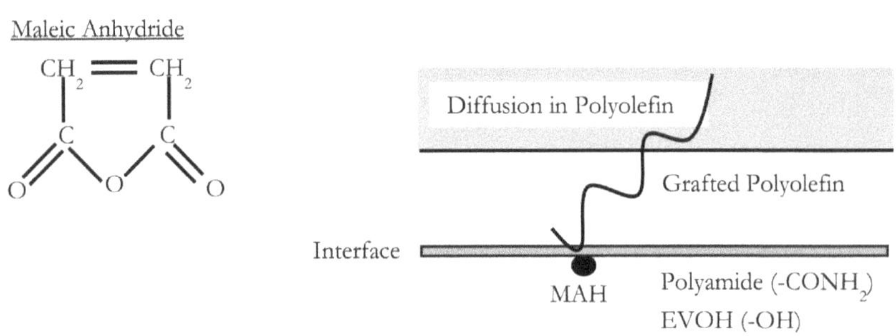

Figure 7.4

Adhesion process at interface with MAH-grafted polymer

the linear PEs is believed to facilitate the reaction. Commercial products are very often made by using the grafted material as a concentrate, which is diluted in other polymers. LLDPE disperses easily in other polyolefins and is recommended as the carrier resin. The dilution level will affect adhesion. In some cases ethylene propylene (EPDM) elastomer is added to refine dispersion. The dispersion of

EPDM in the continuous phase of the host polymer is significantly enhanced by grafting MAH through its double bond. The quality of the dispersion can have a profound effect on bond strength. EPDM elastomers of high ethylene content and low Mooney viscosity are used. The rubber phase will also increase energy dissipation yielding higher adhesion values in peel tests.

Polypropylene can be blended with MAH-grafted LLDPE using 2-5% EPDM as compatibilizer. Such blends can enable the coextrusion of PP with polyamide. This approach is an alternative to grafting directly onto the PP backbone.

These highly specialized materials have been finding increasing use as compatibilizer to blend waste plastics for recycling purposes. Useful blends of recycled EVOH or polyamide with polyethylenes can be produced by the use of ~5% of MAH-grafted polyolefin as compatibilizer.

These chemically modified polymers are formulated according to the demands of the application. Some examples are given in **Table 7.1**. The LLDPE and HDPE grades may contain 2-5% rubber. To minimize cost the resins are diluted by the processor with LD/LLDPE. LLDPE is the more readily dispersed polymer. The bond strength will vary with dilution. As can be seen in **Table 7.1** the maleic anhydride (MAH) weight percent concentrations are low. For thermoforming applications the MAH level should be in the region of at least 0.2% or higher. The stresses during the forming process can cause delamination at the interface particularly at corners.

Table 7.1

Commercial MAH-grafted polymers

MI (dg/min.)	Composition	VA (%)	MAH (wt.%)
1.7	EVA/LLDPE (85/15%)	3.9	0.10
2.0	EVA/LLDPE (80/20%)	3.7	0.12
3.2	EVA/LLDPE (80/20%)	7.5	0.13
2.0	LD/LLDPE (60/40%)	-	0.31
1.5	LDPE	-	0.27
1.1	LLDPE	-	0.36
3.2	LLDPE	-	0.30
3.5	LLDPE	EPDM mod.	0.24
0.85	HDPE	Rubber mod.	0.25

Tables 7.2 and **7.3** compare adhesion values of blown and cast films with PA6 (polyamide) and EVOH. The experiments were based on 5-layer coextrusions with LD/LLDPE blends as the support polymer. The results show significant

differences between the two film processes. The differences between the processes are more marked with the polyamide (PA6) samples than the EVOH. Quenching can be very rapid in the cast film process, which has a profound effect on crystallization particularly polyamide that may affect the interaction at the interfaces.

Table 7.2

Adhesion of MAH-grafted PE to polyamide 6

Base resin	MAH (wt.%)	Bond strength (N/15 mm)	
		Blown film	Cast film
*LLDPE rubber mod.	0.24	10.0	1.5
LLDPE	0.36	Inseparable	0.75
EVA/LLDPE 80/20	0.13	0.96	Inseparable
EVA (8% VA)	0.09	0.2	0.17
EVA/LLDPE 80/20	0.12	0.72	0.55

Diluted to 10%.

Table 7.3

Adhesion of MAH-grafted PE to EVOH

Base resin	MAH (wt.%)	Bond strength (N/15 mm)	
		Blown film	Cast film
LD/LLDPE (60/40%)	0.31	9.70	7.22
LDPE	0.27	8.06	2.25
LLDPE/EPDM mod.	0.30	0.18	0.49

Chapter 8

Extruder design principles

Extrusion is the key process by which all polyolefin films are produced. This chapter will, therefore, review the principals of extrusion with polyethylenes and the machine design features that must be considered to maximize production and film quality.

In any flow channel the melt will use as much of the channel area as it needs. If the melt is flowing too slowly as in **Figure 8.1**, part of the material will build-up and stagnate along the walls. This effect is normally referred to as channeling. Self "scrubbing" from the walls will not occur if the shear stress at walls is too low.

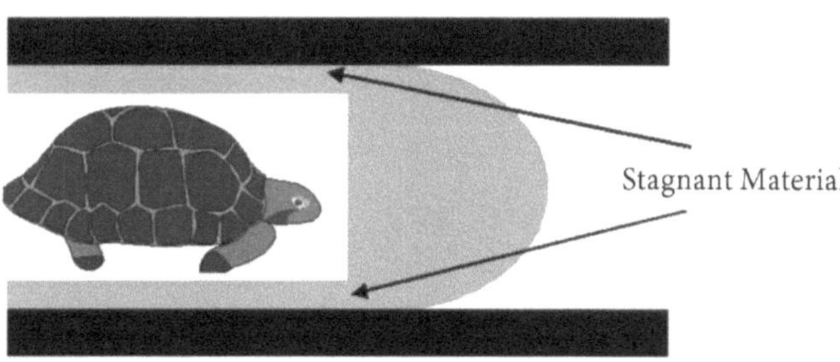

Figure 8.1

Melt stagnation at walls

Machine designers will usually size pipes and other channels according to specified shear rates rather than calculating the shear stress, which actually determines the scrubbing action at the walls. The apparent shear rate is calculated from the velocity and geometry of the flow domain. The equations below show how to calculate shear rates for two simple flow fields.

Apparent shear rate (γ) at the lips of a flat die is calculated as in equation 8.1 where Q is the volumetric mass flow rate, W is the width and t is the die gap.

$$\gamma = \frac{6Q}{t^2(W+t)} \qquad (8.1)$$

The apparent shear rate for a cylindrical section is calculated from the radius (R) as in equation 8.2.

$$\gamma = \frac{4Q}{\pi R^3} \quad (8.2)$$

Similar equations are available for annular dies. Some rules of thumb are used to estimate the minimum shear rates that will avoid channeling and polymer degradation along the walls. For polyethylene the shear rate should be in the region of 20 s^{-1}. With heat-sensitive polymers it is important that wall scrubbing is effective and residence time in the flow domain is minimized.

Figure 8.2 plots the shear stress developed in a pipe 400 mm in length. The mass flow rate through the pipe is 500 kg/hr. The pipe diameter is varied from 10 to 25 mm. The effect of the diameter on shear stress is calculated for an LDPE and an EVOH. Temperatures are 310 °C for the LDPE and 250 °C for the EVOH. As the diameter increases shear stress drops as expected. The EVOH flowing into the 10 mm diameter pipe develops a shear stress of 200 kPa and is likely to develop some surface defects. A larger diameter channel is recommended in this example. In contrast, the LDPE develops a much lower shear stress of 76 kPa in the same geometry. Note that as the pipe diameter increases the curves begin to merge.

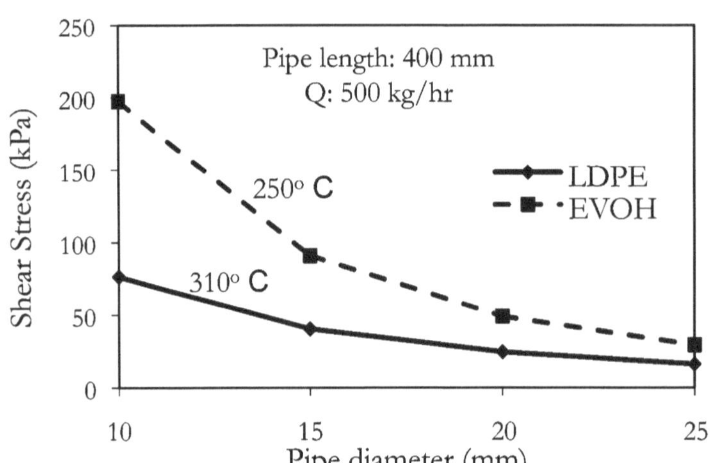

Figure 8.2

Effect of pipe diameter on shear stress

This simple example demonstrates the importance of the flow characteristics and processing temperatures of the materials that are used when configuring the geometry of all the flow paths. Excessive pressures and shear stresses must be avoided but conversely low shear / low velocity conditions will allow stagnation to occur at the walls of the flow domain. This becomes more critical with heat-sensitive and potentially corrosive polymers. Costly polymers such as EVOH will usually be applied at very low thickness. Throughputs will, therefore, be low and flow fields should be designed to develop adequate shear stress to avoid wall stagnation. Problems like gels can form in regions such as the adaptor, connecting pipes, feedblocks and edges of flat dies.

Sharp bends and flat entries into pipes and adapters must also be avoided. **Figure 8.3** shows computer generated flow paths into two pipes with bends. The streamlines show stagnation at all the right-angled bends (ΔP = pressure drop). This is avoided by designing curved bends and tapered entries into narrower channels as shown in the second pipe (no stagnation).

Effective purging of the machine is a necessary part of the operation. It is important to remove residual resin, colour and degraded material. The purging resin must provide maximum scouring and scrubbing action to push out stagnant material from the equipment. The purge resin should be of higher viscosity than the material being cleaned. Melt viscosity, however, should not exceed the maximum pressures and motor loads of the extruder and die. The purge blend should contain fillers such as silica, talc and calcium carbonate to aid in the scrubbing process. To improve penetrating power a small amount of water can be added to produce foaming, which will help the cleansing action.

Materials such as polyamide, EVOH and colour masterbatches can be very difficult to remove. MAH-grafted PE can be added to the purge resin to help in

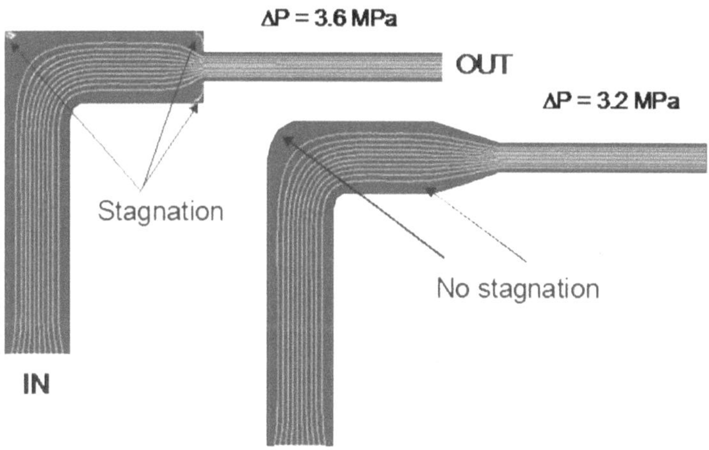

Figure 8.3

Melt stagnation at bends

scrubbing these polymers. It will be necessary to replace the screenpack and clean the die lips and the moveable diverters in coextrusion feed blocks.

An extruder will reach stable conditions from start-up faster if the mass of the barrel and screw are minimized. Most importantly, the smaller the screw diameter is, the shorter the time taken for the machine to reach stable temperature. Extruder diameter has a greater effect on the time to reach stable conditions than length, i.e. the length to diameter ratio (LD ratio). In other words avoid the use of unnecessarily large diameter extruders.

These obvious principals may be overstated here, but they are often neglected when processors are confronted with malfunctioning machinery and tend to expend much effort diagnosing for more complex solutions.

8.1 Single screw extruders

The function of a single screw extruder is to melt the polymer and deliver it at uniform temperature and controlled flow rate into the die, which will form it to the required shape.

Figure 8.4 shows schematically the polymer flow in a single screw extruder. The solids are fed from the hopper into the extruder and progressively heated and melted by the barrel heaters and shear stress generated by the rotational motion of the screw. The molten polymer is then discharged into the die via a connecting adaptor.

Figure 8.4

Melting process in an extruder

Heating is by electrical elements fitted to the barrel and independently controlled. Most extruders will have at least five separate PID (proportional-integral-derivative) controlled heating zones. The thermocouple tips should be placed close to the internal walls. The extruder and die heating system should be able to reach process temperature within the hour. Air cooling of the barrel is also required with blown film lines to avoid overheating particularly with high shear viscosity polymers such as LLDPE.

Figure 8.5 shows a basic screw design layout. The screw is usually split into three zones. The feed zone conveys the solids to the compression zone, which melts the polymer by heat from the barrel and the shear action generated by the motion of the screw relative to the barrel. A molten film is formed, which builds a melt pool in the upstream direction as the channel depth in the compression section is progressively reduced. The fully melted homogenized material is then delivered to the metering zone, where it is pumped at constant pressure to the discharge end of the machine. A mixing (homogenization) section is optional, but highly recommended.

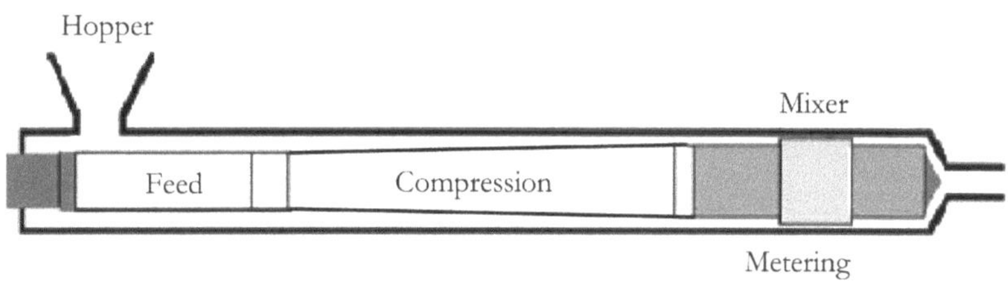

Figure 8.5

Basic extruder screw design

The homogenization process combines two actions; dispersive and distributive mixing. Dispersive mixing is the action of de-agglomeration and breaking down of solid particles and residual unmelted material. Distributive mixing produces random spatial orientation that delivers the melt at constant temperature and consistency. Mixing heads such as the fluted or spiral Maddock, pineapple, pins, blister, etc. designs are used to enhance homogenization. Some designers will include two mixing sections on one screw.

Figure 8.6 shows a basic extruder schematically. The geometry of a screw section is in **Figure 8.7**. Melt is conveyed along the screw (z) and the important dimensions for design consideration are listed in **Table 8.1**. The flight angle depends on the ratio of the pitch to the diameter of the screw. (Screw pitch is the distance between two consecutive flights.) In a square pitched screw the pitch is equivalent to its diameter.

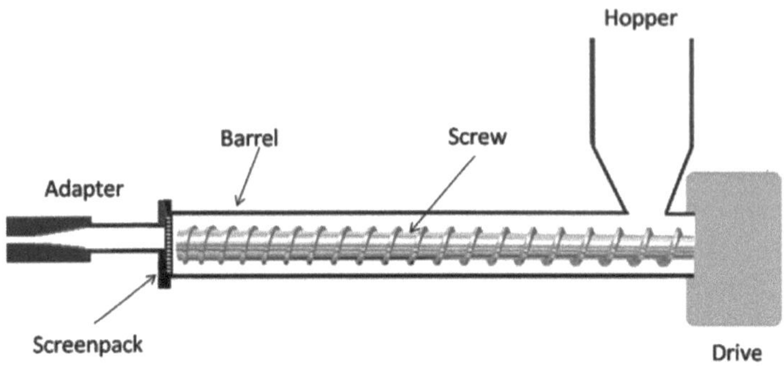

Figure 8.6

Components of a basic extruder

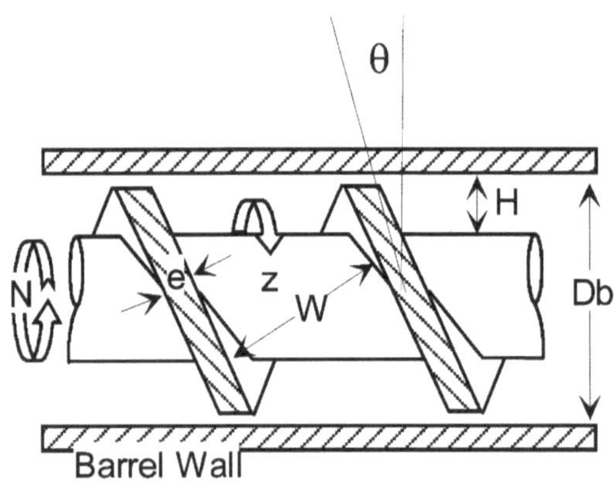

Figure 8.7

Basic single screw geometry

Table 8.1

Key screw dimensions

H	Distance from the root of the screw to the barrel wall
θ	Flight angle (17.66° for square pitched screw)
W	Channel width
e	Flight thickness (usually ~10% of diameter)
Db	Barrel diameter (determines output)
N	Rotational velocity of the screw: determines output (Q)

In **Figure 8.8** the pellets are conveyed into the compression zone where dispersive mixing takes place. Here, the solid and melt phases coexist (Tadmor model). The two phases are divided by the molten polymer accumulating in a melt pool at the rear of the flight and the solids are aggregated as a solid bed at the front of the flight as shown. A thin film is formed between the barrel surface and the solid bed. The contact with the heated barrel and the very high shear in this zone is where most of the melting occurs.

The barrel motion relative to the solid bed drags the melted film into the melt pool. The volume of the melt pool gradually increases in the down channel direction. It is important that the polymer is completely melted before it enters the metering section, which acts as a positive displacement pump.

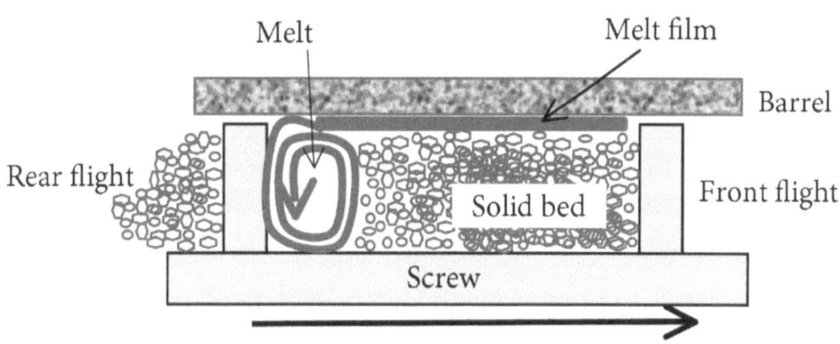

Figure 8.8

Polymer melting in the compression zone (Tadmor model)

A screenpack is placed at the exit to filter out impurities and induce some back mixing. The screenpack should be set up by placing the coarser screens upstream to catch large dirt particles. The finer screens are inserted progressively downstream. A typical breaker plate with screenpack is shown in **Figure 8.9**.

In addition to filtering the screenpack converts rotational memory into longitudinal memory. Lever or ratchet operated automatic changers are used to replace contaminated screenpacks without machine stoppage.

A well designed screw should be free of surging or output pulsations. High level pulsations can be picked up with pressure gauges and ammeters. Low level pulsations are more difficult to detect as pressure transducers may not measure real levels of melt flow fluctuations. The best method of detecting low level pulsations is to continuously monitor thickness variations on-line.

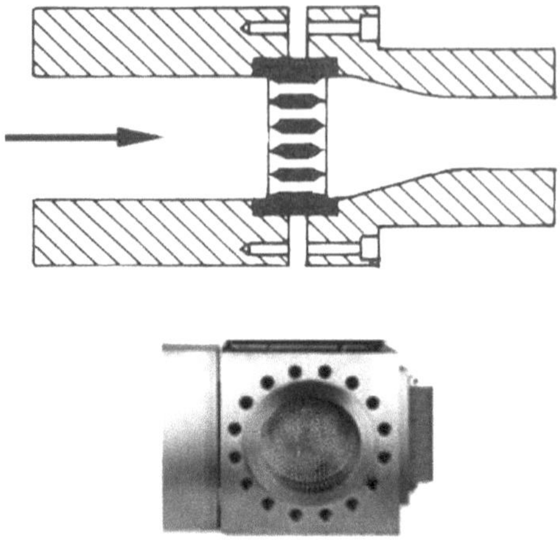

Figure 8.9

Breaker plate and screenpack

The following factors can all be a source of surging:

- Drive variations
- Gear slippage
- Worn screws
- Malfunctioning heaters
- Burned out fuses
- Corroded thermocouples

All the above will contribute to surging or pulsations as well as poor temperature control, which is crucial for thickness distribution. The key machine design features that can influence the performance of the extruder in this respect are as follows:

- Feed hopper
- Solids transport (feed zone)
- Melting zone (compression zone)
- Melt conveying (metering zone)

The process starts with the flow of pellets from the hopper. **Figure 8.10** shows a simple hopper feed section. It is important that the slope and internal surface of the hopper allows the polymer pellets to flow freely and evenly into the throat and feed zone of the screw. Pellets should not funnel or arch at or around the throat resulting in uneven material flow. The polymer silo storage and piping to the extruder hopper must be regularly checked for any build-up of fines, angel hair and dust, which may hinder the flow of pellets and cause cross contamination and form gels in the melt.

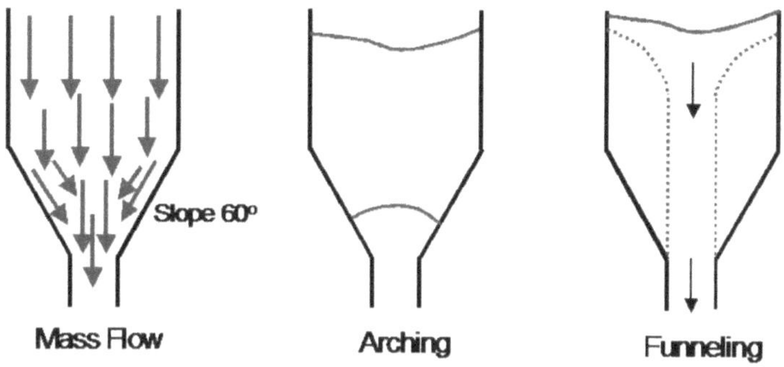

Figure 8.10

Pellet flow within the hopper

In **Figure 8.11** the material must flow freely into the throat of the hopper and be carried away by the screw without melting, agglomerating or sticking. The pellets must be of uniform size and cut and dust-free. Most extruders include a metal feed casing with the opening and the throat above it, as well as the surroundings for the first few flights of the screw. The casing must have water passages that cool all these areas. A temperature of 30-40 °C must be maintained in this zone. When extruding low melting polymers some designers recommend cooling the

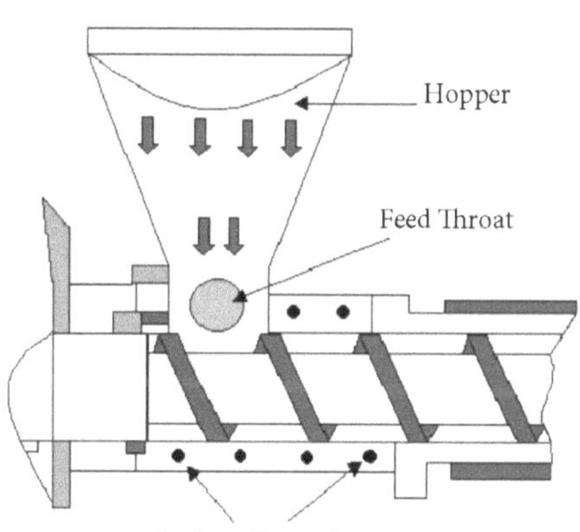

Figure 8.11

Cooling at the feed throat

screw root over the first few flights. Normally screw cooling is not recommended as it will reduce extruder output and waste energy.

An important source of extruder surging originates in the compression zone. Solid bed wedging and solid bed break-up are two phenomena that can occur in this zone, which will lead to various degrees of surging. Both phenomena originate from a poorly designed compression section. The compression zone must allow enough time for complete melting of the solids.

Figure 8.12 illustrates the formation of solid bed wedging. This occurs when the screw channel volume is reduced too rapidly. Not enough time is available to completely melt the material and a proportion of the solids are wedged at the front of the flight leading to surging.

In **Figure 8.13**, again due to a poorly designed compression zone unmelts may be formed as the solid bed breaks up. These unmelts swim in the melt pool, where shear stress is low, and will cause pulsation as they flow through the discharge end of the extruder.

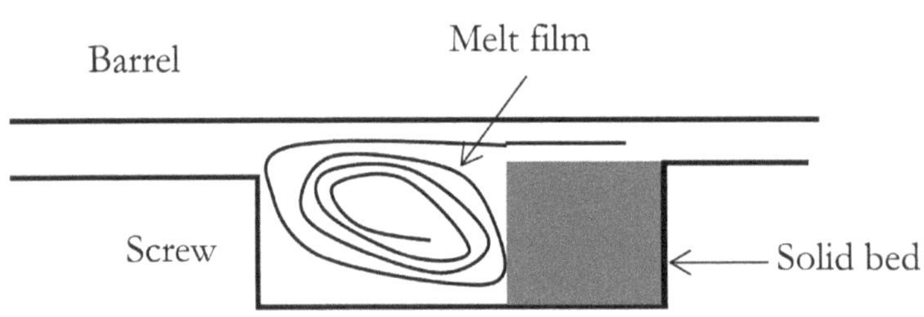

Figure 8.12

Solid bed wedging

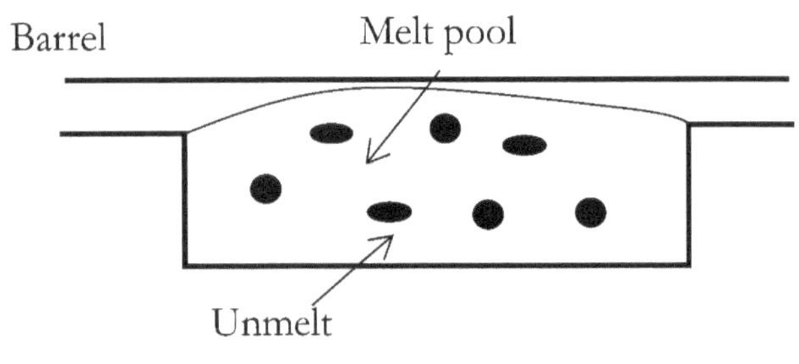

Figure 8.13

Solid bed break-up

A computerized plot of the melting process is shown in **Figure 8.14**. The solid bed ratio shows the progression of the melting process. The solids gradually diminish as the material flows along the screw. The LDPE (0.8 dg/min. melt index) is fully melted at around 15 D along the screw. The simulation shows that only fully melted polymer is entering the metering section in the final third of the screw.

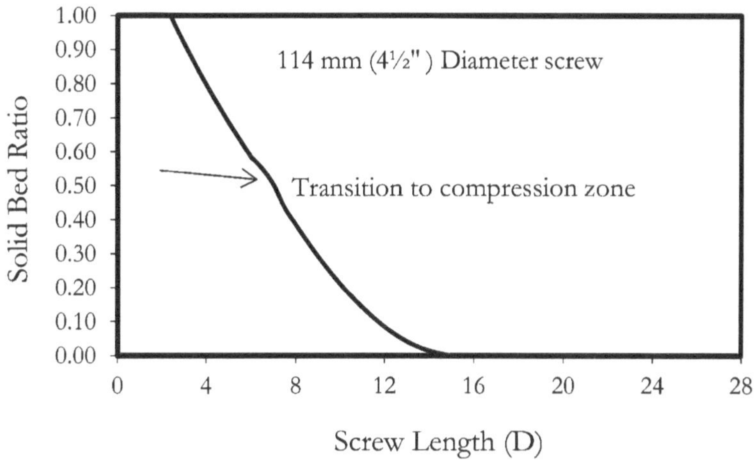

Figure 8.14

Solid bed melting in metering screw
Temperature: 230 °C

The alternative to the compression melting process is the barrier design. A barrier screw includes a double flighted zone, which separates the melt from the solids and ensures that only fully melted material reaches the metering section. By containing the solids in the primary (solids) channel maximum contact is made with the barrel wall where melting occurs. The molten polymer flows in the secondary channel to form a melt pool as shown in **Figure 8.15**. The barrier gap created by the flight height separating the two channels is critical.

A correctly designed barrier screw should ensure that the primary channel is always full with solids. Two basic designs are used. The principle of the Hartig Barr design is to progressively reduce the depth of the primary channel held at constant width while the melt channel depth increases. This offers the maximum surface for the shearing action of the solids on the barrel to occur. The alternative is based on the Maillefer travelling barrier design, which continuously reduces the solids channel width at a constant depth. Combinations of the two concepts have been designed.

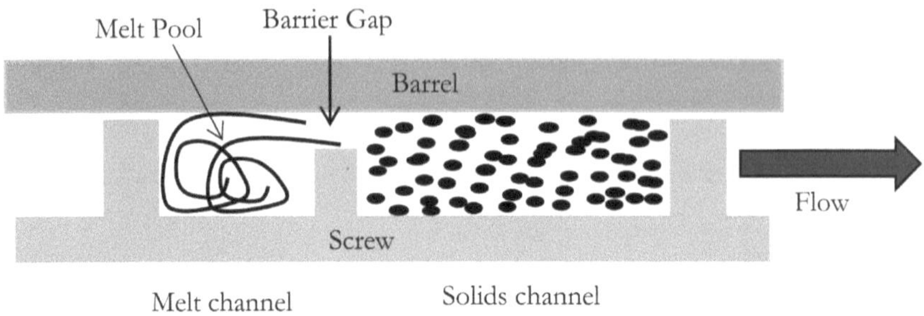

Figure 8.15

Principle of separation in a barrier screw (based on Tadmor model)

Figure 8.16 compares the melting process of LDPE and mLLDPE in a 90 mm diameter, 24:1 LD ratio barrier screw. The feed zone has a channel depth of 12.0 m with a deep metering channel of 8.0 mm (low compression). The barrier section has a pitch of 1.3 D and the secondary flight gap is 1.27 mm. The melting process as defined by the solid bed ratio shows both polymers melting at the same rate and flowing into the contracting (reducing depth) primary channel, which is filled with solids from 11 D to 18 D. In this zone the melting rate is maximized as the contracting channel depth stays full of material. At around 18 D the channel is completely depleted of solids as the melt has filled the expanding secondary channel and flows into the final metering zone. Both polymers melt at approximately the same rate. The exit temperature of the mLLDPE is 16 °C higher than the LDPE at 100 rpm.

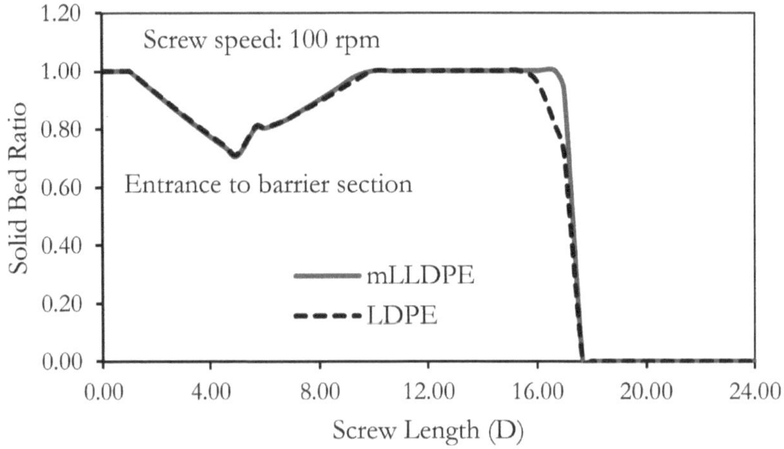

Figure 8.16

Melting in primary channel of barrier screw

Barrier sections are more costly to construct than metering screws. However, their claimed higher output and greater energy efficiency has won them many adherents. Barrier screws are also used with polyamide and polyesters. Smooth barrel feed sections are recommended with these materials.

8.2 Mixing sections

Distributive mixing should occur downstream to deliver a fully homogenized melt into the final discharge section. Some form of proven mixing section is advisable particularly when blends and recyclates are to be processed. Mixers such as the Maddock are best placed no more than 2 D from the end of the screw. Some designers recommend only a half turn away from the screw tip.

A sketch of a Maddock fluted mixing section is shown in **Figure 8.17**. The melt is split and feeds into channels. Homogenization is achieved by the circular motion of the melt in the channel, which then leaks over the gap into an adjacent channel where further mixing occurs. Four to eight channel pairs are usually recommended to optimize mixing with this design. The mixing is mainly dispersive.

The effectiveness of a Maddock mixing section was tested by computer simulation. The simulations were based on a 90 mm diameter screw running with an octene LLDPE (MI=1.0 dg/min.) The Maddock is a 5-channel pair with a 1.3 mm clearance from the barrel surface. The screw metering section is 8.2 mm deep.

In **Figure 8.18**, the melt temperature cross channel variation (ΔT) just prior to entry into and exit from the mixer is shown. The melt temperature in the screw

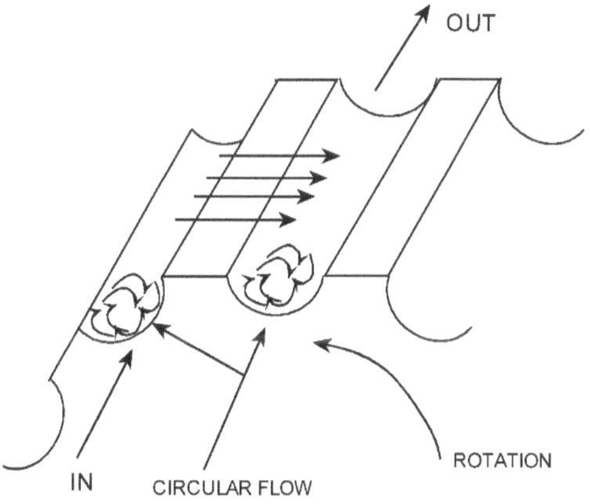

Figure 8.17

Melt flow within a Maddock mixing section

channel varies (in) from 210 to 225 °C. The mixer (out) both increases the temperature and flattens its distribution (ΔT) illustrating its mixing capability. The exit temperature is now much more uniform at ~235 °C. In this example, the clearance between the mixer and the barrel should be increased to reduce overheating.

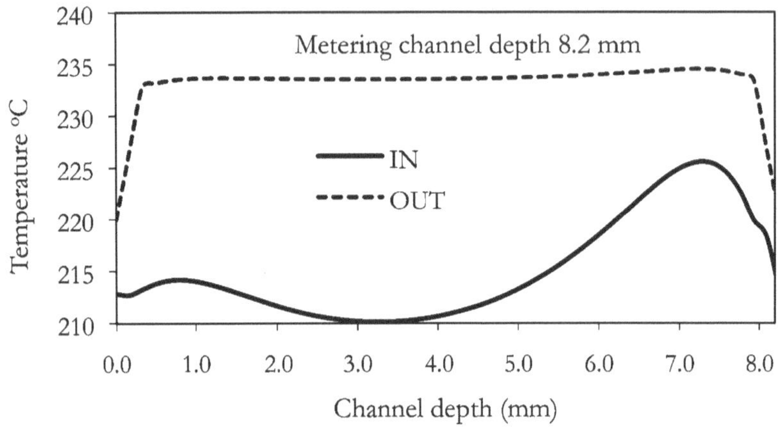

Figure 8.18

Temperature variation in melt channel of mixing section

Other mixer options are pineapple and pin designs. These develop distributive mixing and can be combined with a Maddock section. **Figure 8.19** shows a typical pin mixer.

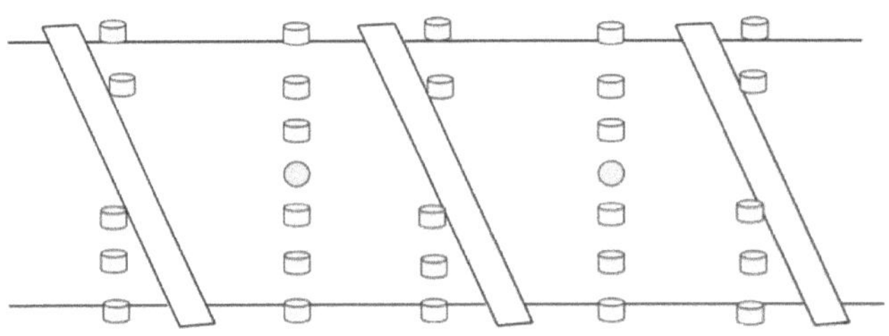

Figure 8.19

Principle of a pin mixer

If a large melt temperature variation is allowed to flow into the die film thickness becomes more difficult to control. Substantial temperature variations have been measured with variable depth thermocouples.

The proliferation in the use of polyethylene blends and the inclusion of scrap and recycled materials makes it imperative to use the most efficient design of mixer to produce films of acceptable quality and consistency. High LD ratio screws are recommended with the appropriate mixer design.

8.3 Extruders with grooved feed sections

Grooved barrel feed sections are commonly used as a means to boost output from any given size of extruder. The concept was originally developed for HMW-HDPE as a method to improve control of melt temperature. **Figure 8.20** shows a typical rectangular grooved section.

If the barrel wall in the feed section is smooth, the polymer melt conveyed by the screw will depend on the friction ratio and consequently on temperature. If the barrel is grooved, its geometry will govern the friction ratio. The barrel grooves may be rectangular, hemispherical or helical in cross section and usually run in the axial direction. Grooved sections must, however, be intensely cooled to prevent the polymer from adhering to the barrel wall.

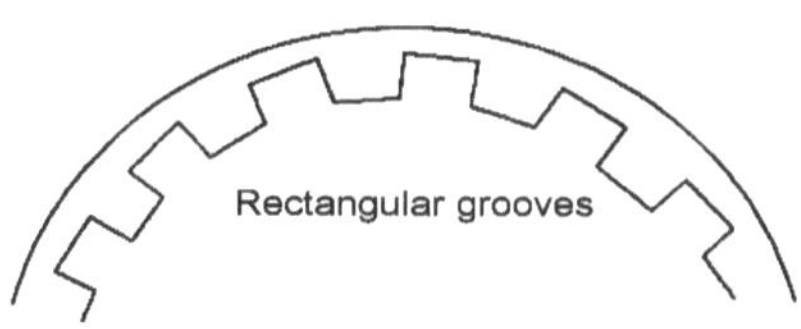

Figure 8.20

Typical grooved barrel section

The pellets feeding into the barrel are prevented from rotating with the screw since they are firmly grasped in the plane of intersection between the barrel surface and the grooves. This ensures more efficient conveying along the barrel by the screw than would be the case in a temperature controlled smooth feed section. The channels formed by the grooves will taper off conically towards the end of the feed zone thus compressing the polymer as the volume is reduced. The net effect is to increase frictional heating, which melts the polymer as it reaches the end of the grooves. It is important to provide some insulation between the cooled feed zone and the heated barrel. A replaceable liner can act as insulation and will also avoid undue wear of the barrel. The harder the pellets the easier is their conveyance along the barrel. As a result the higher the PE density the greater the volume displaced per revolution along the screw. A recommendation with grooved feeds is to design the feeding zone with a shallower channel depth than normal.

Table 8.2 shows typical improvements in conveying of the grooved feed over the smooth feed.

Table 8.2

Feed conveying capacity

Groove type	Specific feed rate (kg/hr/rpm)
Square pitch 63 mm screw	
None	1.52
Parallel groove	1.82
Helical (clockwise)	1.89

In **Table 8.3** the results obtained with a decreasing pitch screw designed for a smooth barrel extruder was compared in both a smooth and grooved barrel machine. The extruder had a 24:1 LD ratio. The LLDPE was of 1.0 dg/min melt index processed at 100 rpm screw speed.

Table 8.3

Output comparison for grooved versus smooth feed section

	Smooth	Grooved
Output (kg/hr)	100	116
Melt temp. (°C)	218	204
Torque (kW/rpm)	0.24	0.34

The grooved feed increases output by 16% and reduces temperature. However, torque increases. Screw design modifications are necessary to reduce torque without sacrifice in output, e.g. shorter barrels. Machinery suppliers offer proprietary screw designs based on grooved feed sections that are claimed to optimize output and minimize screw wear with LLDPE and HMW-HDPE.

Some researchers have claimed that the conventional smooth bore extruder has clear advantages in film quality and film properties over the grooved feed extruder for LLDPE film production. All the following properties are claimed to improve with the smooth bore feed:

- Dart impact strength
- MD tear resistance
- Puncture resistance
- Lower gel level in film

However a grooved feed screw offers the following advantages:

- Extrusion stability (less surging)
- Better melt temperature control at high outputs

8.4 Screw and cylinder wear

Since the introduction of LLDPE the incidence of screw and barrel wear has increased considerably. The reasons are largely due to the reduced shear sensitivity of LLDPE compared to LDPE. At the shear rates encountered in typical extrusion operations LLDPE will have a higher viscosity for the same melt index and extrusion temperature. This leads to higher pressure and friction, which increases the rate of wear on screw and barrel surfaces.

As far as operating conditions are concerned, there is little that can be done without negative side effects. Lower screw speeds will reduce wear, but will obviously reduce output proportionally. Another way of reaching the same result is to use larger sized extruders to deliver the same output with lower screw speeds.

LLDPE grades with a higher melt index will reduce wear, probably explaining the better experience with cast film extruders, where higher melt index grades are used at higher temperatures. Processing aids (e.g. fluoroelastomers) reduce pressure and could improve wear resistance.

The ideal solution is to develop linear PEs with broader MWD via cascade polymerization processes and/or more versatile catalyst systems. Research to produce bimodal MWD distributions and constrained geometry metallocene catalysts, which induce long chain branching, are ongoing amongst polyethylene producers.

8.5 Extrusion of metallocene-catalyzed LLDPE

In a study using computer simulations a commercial mLLDPE is evaluated in an extruder designed for blown film processes. The extruder has a 90 mm diameter and is sketched in **Figure 8.21**. To minimize pressure a low compression barrier screw is entered. The hexene mLLDPE has a melt index of 1.0 dg/min and a density of 0.918 g/cm.

The barrier gap is 1.27 mm and the clearance in the fluted mixer is 1.3 mm. The metering zone depth is 8.3 mm. The fluted mixer is a Maddock device with five channel pairs. The pitch in the barrier section is 1.3 D and standardized to 1 D in the feed and metering zones. The LD ratio is 24:1. The recommended barrel temperatures are shown in **Figure 8.21**. These as can be seen are set low (185 °C) to avoid excessive melt temperatures.

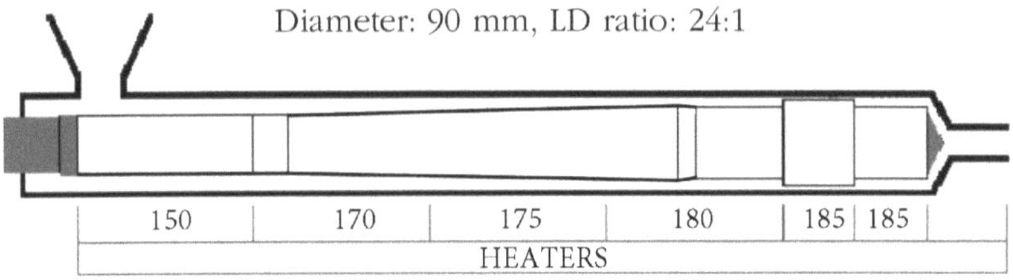

Figure 8.21

Schematic of barrier screw

The results of these virtual experiments are summarized in **Table 8.4**. Over the screw speed range from 30 to 100 rpm the melt temperature rises from 203 to 239 °C. This is despite the low barrel settings of 180-185 °C in the metering zone. There is an overshoot of some 50 °C at the higher outputs. The peak pressure at the entry of the barrier zone increases from 62 to 80 MPa over this output range. The length of the melting zone is consistent throughout.

Table 8.4

Virtual extruder output data with mLLDPE

Screw speed (rpm)	Output (kg/hr)	Exit temp. (°C)	Peak pressure (MPa)	Melt zone (D)
30	93	203	62	17.5
50	134	215	64	17.5
60	154	221	73	17.5
75	184	228	76	17.5
100	233	239	80	17.5

The solid bed ratio in the primary channel is plotted in **Figure 8.22** at a screw speed of 100 rpm. The solids ratio drops as the material begins to melt as it enters the barrier section with its dual channel. By 18 D, the solids in this channel are completely depleted as the fully melted material flows into the metering zone from the secondary channel. The simulation shows that the primary channel is full with solids from 9-18 D as its volume is continually contracting and the expanding secondary channel is filling with melt.

These virtual results are compared to actual experimental runs in **Figure 8.23**. Correlation between the two sets of data is very good. At the highest output (screw speed 100 rpm) the simulated temperature is slightly lower.

The same screw and temperature profile was simulated with a film grade LDPE (0.8 dg/min. MI). The results are compared to the mLLDPE in **Figure 8.24**. At a screw speed of 100 rpm, the LDPE develops an exit temperature of 223 °C compared to 239 °C for the metallocene.

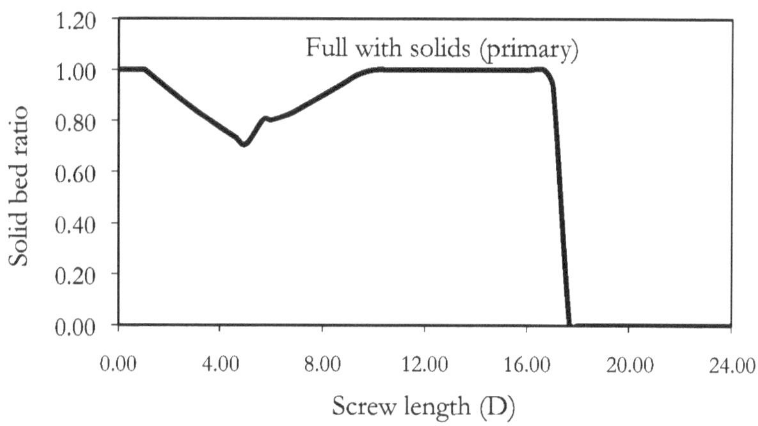

Figure 8.22

Melting of mLLDPE along barrier screw

Screw Speed: 100 rpm

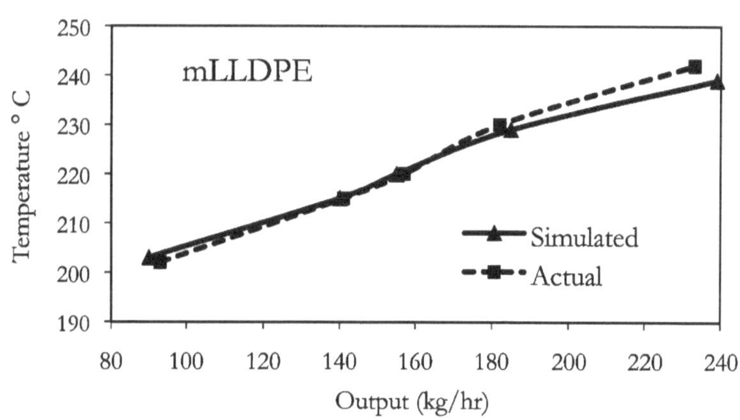

Figure 8.23

Actual versus simulated output data

In the next experiment the mLLDPE is compared to a Z-N LLDPE with properties shown in **Table 8.5**. The barrel temperature settings in the metering zone were reduced by 10 °C (175 °C).

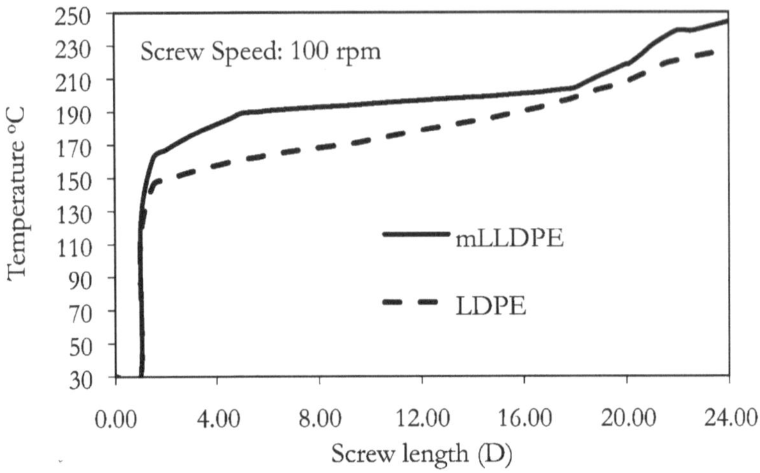

Figure 8.24

Temperature profile of mLLDPE compared with LDPE

The shear viscosity curves for the two resins (calculated using the Cross model) are plotted in **Figure 8.25**.

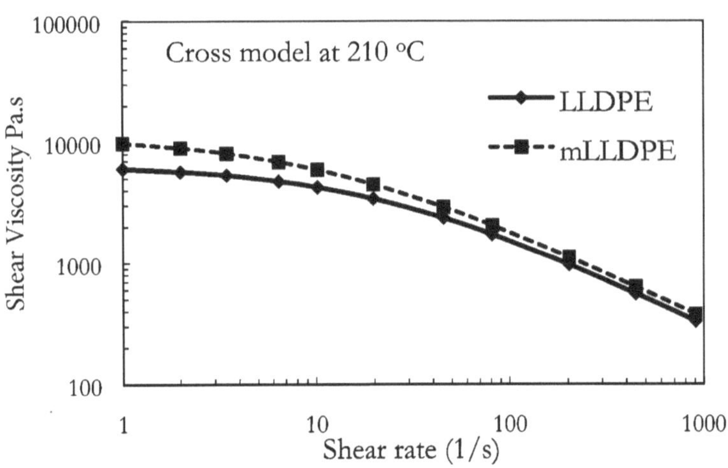

Figure 8.25

Shear viscosity of LLDPE and mLLDPE

Extruder output data for the two resins are shown in **Table 8.6** at a screw speed of 75 rpm. The mLLDPE exit temperature is 220 °C and the LLDPE is 216 °C. Output per screw revolution is increased with the higher viscosity metallocene polymer.

Table 8.5

Basic polymer characteristics

Polymer	Melt index (dg/min)	Density (g/cm^3)
LLDPE (n-butene)	1.0	0.918
mLLDPE (n-hexene)	1.0	0.918

Table 8.6

LLDPE output compared with mLLDPE

75 RPM screw speed	LLDPE	mLLDPE
Output (kg/hr)	168	178
Output/rpm (kg/hr)	2.24	2.37
Exit temp. (°C)	216	220
Power (kW)	60.5	68.6

The conclusions from these evaluations areas follows:

- At high outputs the linear polymers overshoot the barrel temperature settings by around 50°C with this barrier screw design.
- The screw is relatively short at 24:1 LD ratio. In film processes longer screws are preferred to optimize flow stability.
- The mLLDPE although requiring more power does deliver more output per screw revolution.
- Overall, the differences in processability between the conventional LLDPE and the metallocene LLDPE were not large.
- A processor familiar with LLDPE extrusion should have few problems processing mLLDPE.

Finally, excessive viscous dissipation (frictional energy) as observed here is best avoided as this may increase the tendency of melt surging from the machine. Ideally, the process should be run close to isothermal conditions. In these evaluations the process is operating under a mainly adiabatic regimen.

Chapter 9

Tubular film production

The underlying principle in die design is to avoid cross flows, which lead to uneven flow rates across the exit channel. A correctly designed die should meet the following basic criteria:

- Balance the flow for uniform thickness distribution
- Maintain temperature homogeneity
- Maintain a reasonable pressure drop
- No hot spots
- No stagnant flow
- Avoid die lip build-up
- Avoid sharkskin and melt fracture
- Minimize weld lines
- Minimize residence time (shear rates >10 s^{-1})

The function of the tubular die is to form the melt into an annular shape with evenly distributed flow round its circumference. The basic design is a hollow cylindrical heated body with a concentric supported mandrel insert. Various methods to support the mandrel were developed over the years. The widely used spider systems work well. However, the melt stream is invariably split, by the supports (legs), leaving weld lines in the machine direction (MD). Smear devices are usually introduced to minimize the weld lines. These divert the melt at the inner wall into the peripheral direction with the intention to eliminate flow lines in the axial direction along the outer wall. However, the thickness profile will still retain the flow lines, albeit reduced, made by both the spider legs and the smear device. These may still reduce the tear resistance and impact strength of the film.

The more complex spiral mandrel die is a more effective way of forming an annular section with even melt flow distribution round its circumference. A schematic of a typical spiral mandrel section is shown in **Figure 9.1**.

The melt from the extruder flows into a series of ports connected to the channels, which are machined into helical paths. As the melt moves round the channels pressure builds up and leakage over the channels in the machine direction (MD)

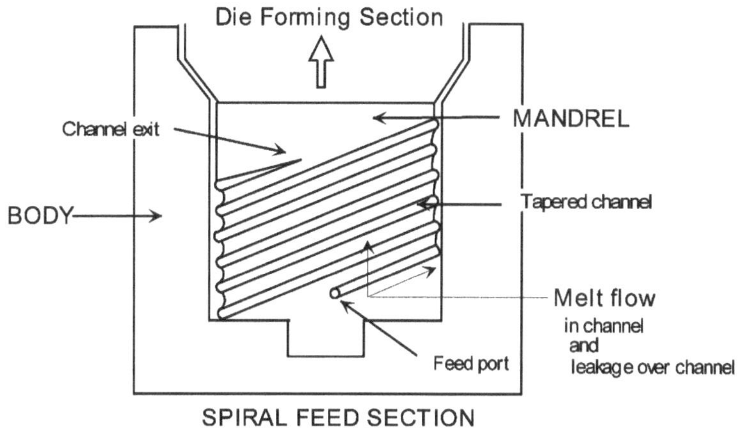

Figure 9.1

Spiral mandrel die section

occurs. The melt stream is thus redistributed along two axis. As the melt exits the spiral section it feeds into the channel leading to the die lips, which form the tubular film.

Critical spiral mandrel design features include the following:

- Number of entry ports (1 per 25 mm of diameter)
- Number of channel overlaps
- Gap between body and mandrel
- Length to diameter ratio
- Depth of spiral channel

High pressure will usually improve melt flow rate distribution. A typical pressure drop with LDPE is 150 bar.

Weld lines are still formed as the melt travels from the outer surface of the annulus to the inner surface via a circular path. However, a layering effect occurs which welds together the polymer from the entry ports. With systems based on spider supports the weld lines are in a plane through the melt that flows through the die axis. These are more likely to produce weaker film.

From the spiral mandrel the evenly distributed melt stream is fed to the outflow channel (die lips). The die gap at the lips has to be adapted to the polymer's rheology and the specified film thickness. If the gap is too narrow melt fracture can occur and if it is too wide drawdown rate becomes too high and melt break will occur. For LDPE, EVAs etc. die gaps of 0.8-1.2 mm are recommended. With LLDPE wider die gaps of 2.0-2.5 mm are usual. HDPE used in blown film will normally have very broad MWD and consequently narrower gaps can be used.

The die gap design variables to be considered are as follows:

- The onset of sharkskin is at a shear stress of 0.14 MPa. Eventually melt fracture will occur as higher shear stresses develop. This is a problem with LLDPE and mLLDPE.
- For good thickness distribution an optimum head pressure is required.
- LLDPE melt has less memory compared with highly branched LDPE resins. This means there is a lesser effect from entry angles, restrictions or final land tapers.
- Shear rates should not be lower than 20 s^{-1}. Minimize residence times particularly with heat-sensitive and corrosive materials.
- Excessively wide die gaps, to reduce shear stress, will lower film properties as film orientation becomes increasingly unbalanced. Thickness distribution and drawdown will also suffer.

Figure 9.2 shows some die exit profiles for LDPE, LLDPE and HDPE. The die gaps for HDPE and LDPE are 1 mm. However, HDPE has a shorter land length

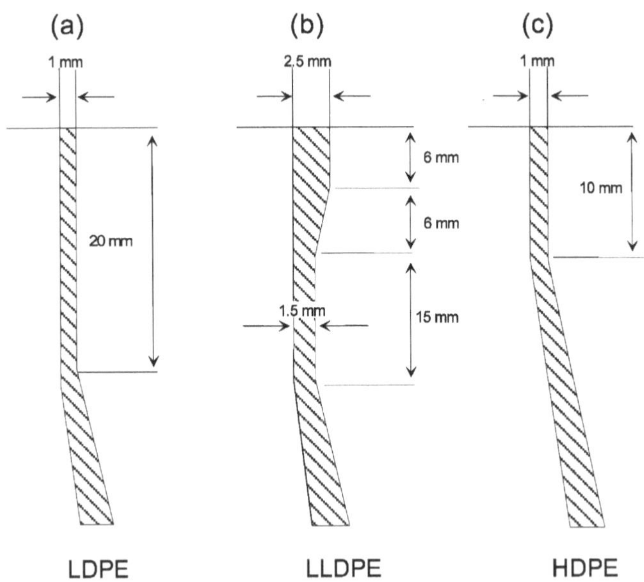

Figure 9.2

Typical die exit flow paths

of 10 mm compared to 20 mm for LDPE. This is to avoid excessive backpressure with HMW-HDPE. The LLDPE is the least shear-sensitive and requires a wider die gap.

Figure 9.3 illustrates fine tuning of the die gap by moving the die ring with a screw mechanism.

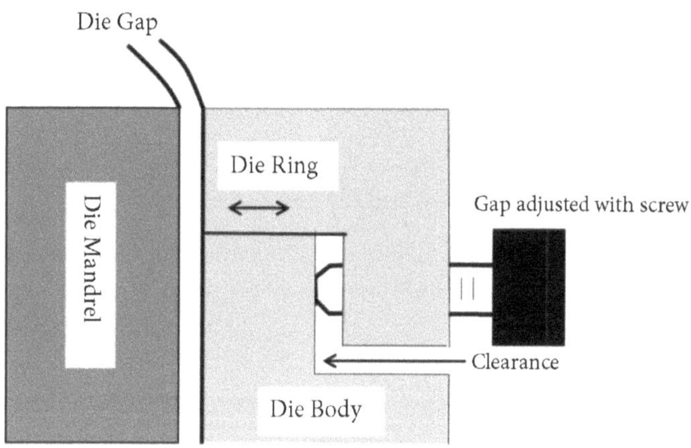

Figure 9.3

Die gap adjustment

9.1 Die build-up

An important consideration in die design is the possible formation of die build-up, which can occur with all types of dies. This is an exit phenomenon, which results in debris from the melt attaching itself to the edge of the die lip and subsequently contaminating the extrudate surface. This will necessitate machine stoppages for cleaning. Die build-up is associated with the following phenomena:

- Die swell
- Broad MWD resins
- Additives
- Oxidation
- Moisture

The presence of pigments, e.g. TiO_2 which may initiate localized polymer degradation can also generate build-up. Pigment masterbatch selection is, therefore, crucially important in minimizing the risk of such phenomena occurring. The following suggestions may all help to reduce die build-up:

- Build-up is more likely to form with sharp edged lips. Therefore a recommended method to minimize this defect is to have rounded edges (radius 6-12°) at the lip.
- Addition of 100-200 ppm fluoropolymer (PPA) can help to minimize die build-up. The explanation for its effectiveness is that it forms a non-adherent layer on the extrudate, which stops the oxidized particles from sticking to its surface.
- Metallurgical treatments such as titanium or boron nitride will induce slip and reduce wall shear stress and die swell.

There is some evidence that mLLDPE (metallocene), with its narrower molecular weight and composition distributions is less prone to die build-up than Z-N LLDPE.

9.2 Design of spiral dies

The three sections of the 350 mm diameter die used in the following evaluations are shown in **Figures 9.4** to **9.6**, a 45 mm diameter entry pipe leads into the 14 ports, which feed the 14 spirals in this design. The use of 14 ports is based on the design principle of 1 port per 25 mm of die diameter (350 mm). The ports are 15 mm in length and 15 mm in diameter.

In **Figure 9.5** the geometry of the spiral section is shown. The body is a cylinder with a 353 mm internal diameter from top to bottom. The mandrel is designed to form a gap of 0.2 mm at the bottom, opening up to 2.5 mm at the top. The spiral channels have a radius of 5 mm and range in depth from 17 mm at the entry tapering off to 0.1 mm after six overlaps. This leads over a 5 mm land to the outlet section shown in **Figure 9.6**.

In **Figure 9.6** the melt from the spirals enter a 5 mm long section with a 2.5 mm gap, which flows into a 30 mm length channel, which compresses the melt from 2.5 mm to 2 mm. A short 5 mm channel reduces the gap to 1 mm, which leads to the final 10 mm land. The final gap was originally set to 1 mm as shown in **Figure 9.6**. This gap was increased to 1.5 and 2.2 mm respectively in separate experiments.

The key element that determines die performance is the spiral section, which forms the melt into an annular shape of even distribution round its circumference. In **Figure 9.7** the volume flow rate distribution from the spiral exit is compared to the flow exiting from the die lips. The output is from one spiral element. It is good design practice to ensure that the bulk of the melt distribution is executed by the spiral section and that the final exit zone carries out the minimum flow rearrangement.

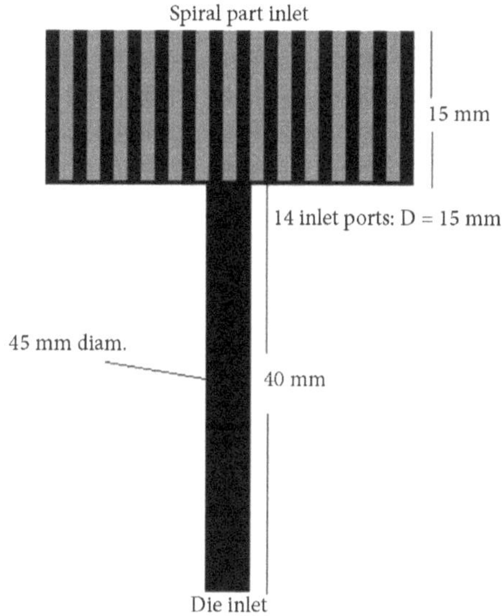

Figure 9.4

Inlet to spiral section

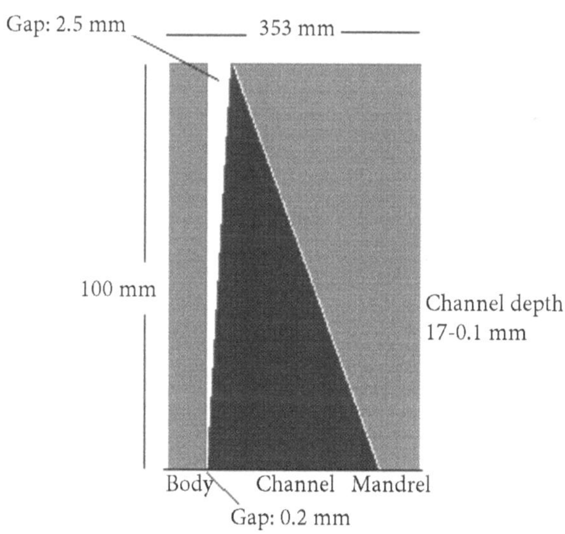

Figure 9.5

Geometry of spiral section

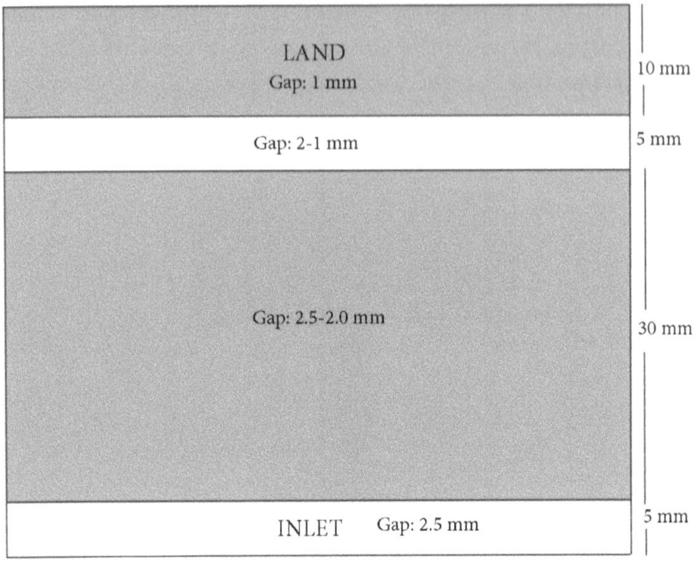

Figure 9.6

Outlet (lips) section

The material exits the spiral section with a thickness distribution of ±2.32% and is reduced ±0.26% past the lips. The linear polyethylenes are less viscoelastic, which facilitates melt distribution in dies. In contrast LDPE is more viscoelastic (higher shear sensitivity) and consequently flow distribution will require more correction.

LLDPE will develop excessive shear stress with this 1 mm die opening.

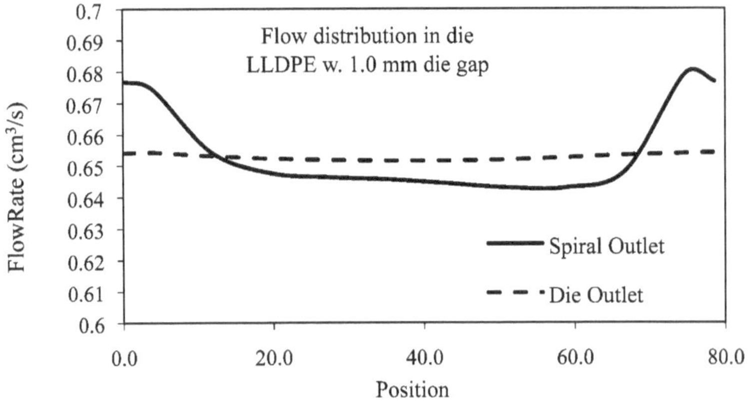

Figure 9.7

Volume flow rate at spiral exit and die lips

9.3 Virtual experiments with the 350 mm spiral die

In this virtual experiment the effects of die gap and polymer selection are compared in the 350 mm diameter die.

All die walls and the melt were set at a temperature of 230 °C. Polymer input was 250 kg/hr (2.27 kg/hr/cm) in all experiments. Three die gaps were evaluated with five polyethylenes recommended for blown film applications. **Table 9.1** compares the temperature rises, pressure drops and shear stress at the three die gaps.

Table 9.1

Summary of spiral die runs

Polymer	Die gaps		
	1.0 mm	1.5 mm	2.2 mm
LLDPE (octene)			
ΔT (°C)	5.0	4.0	3.0
ΔP (kPa)	14.6	11.7	10.3
Shear Stress (kPa)	238	157	101
LLDPE (butene)			
ΔT (°C)	6.0	5.0	4.0
ΔP (kPa)	19	15.3	13.6
Shear Stress (kPa)	275	193	131
mLLDPE (hexene)			
ΔT (°C)	7.0	6.0	5.5
ΔP (kPa)	23	19.6	17.7
Shear Stress (kPa)	307	230	165
mLLDPE (octene)			
ΔT (°C)	5.5	4.0	3.7
ΔP (kPa)	16.8	13.4	11.7
Shear Stress (kPa)	272	180	116
LDPE (tubular/0.8 MI)			
ΔT (°C)	2.0	2.0	2.0
ΔP (kPa)	7.2	6.4	5.9
Shear Stress (kPa)	78	65	52

In **Figure 9.8** the pressure drops for all the materials are compared with the 1.0 and 2.2 mm die gaps. The LDPE develops the lowest pressure as expected and the hexene (C_6) mLLDPE the highest.

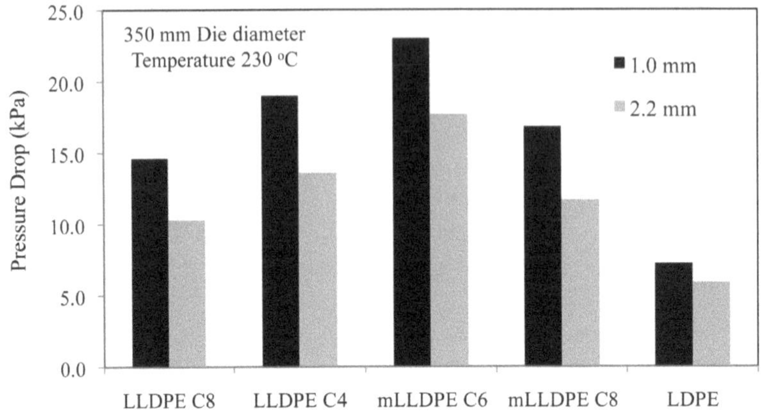

Figure 9.8

Pressure drop for 2 different die gaps

Figure 9.9 shows the effect of die gap change on shear stress at the die exit for all the PEs in the program. Assuming 140 kPa as the threshold for the onset of sharkskin only the LDPE can be run with a 1 mm gap at these process conditions. A 2.2 mm gap is probably safe with all the linear polymers in this series. This prediction shows that the octene LLDPE (C_8) develops the lowest shear stress amongst the linear polymers.

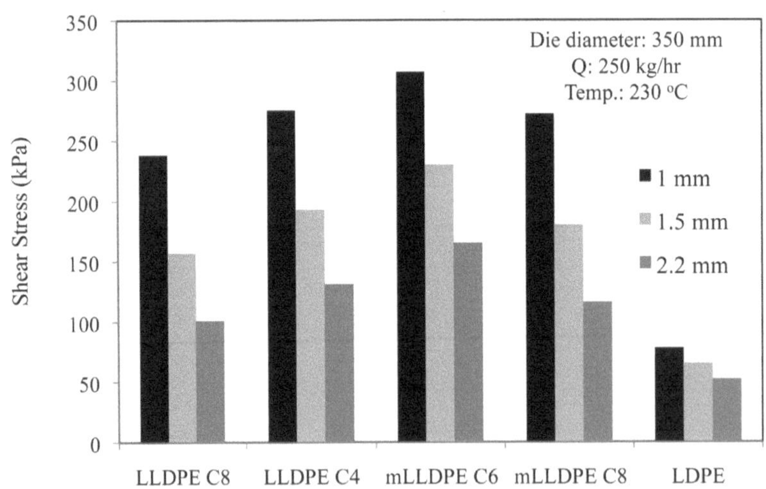

Figure 9.9

Effect of die gap on shear stress

In **Figure 9.10** the temperature increases in the die caused by viscous dissipation of all the materials at the 1.5 mm die gap are shown. The temperature rise is from the set point of 230°C. The LDPE increases by only 2 °C whereas the mLLDPE (C_6) increases by 6 °C. This is accordance with the pressure drop data in **Figure 9.8**. The effect of viscous energy dissipation in the die is often neglected by processors, but as these simulations suggest the temperature rise can be significant with some materials. This data does not include the viscous energy generated in the extruder feeding the die, which can be considerable.

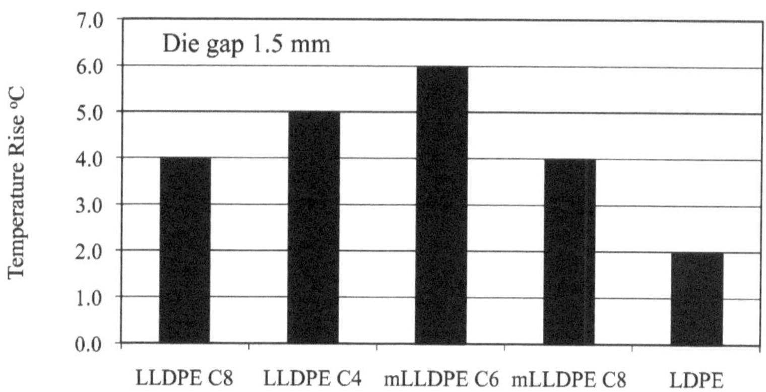

Figure 9.10

Temperature rise in the spiral die

The differences in flow properties between the general purpose LDPE grade and all four linear polymers are clearly contrasted in these simulations. There is also a difference albeit smaller between the metallocenes and the Z-N linear PEs. Wider die gaps (2.0-2.5 mm) are clearly required with the linear polymers to avoid sharkskin at commercial outputs.

9.4 Solutions for melt fracture and sharkskin

Sharkskin and melt fracture occurs when the polymer melt leaves the die above a critical shear stress. The onset of sharkskin with PE is generally agreed to be at a shear stress of 0.14 MPa at the die wall. Melt fracture will occur as the stress approaches 1.0 MPa. The causes are not fully understood, but are related to the following phenomena:

- Normal stresses developed by the polymer

- Die land slippage
- Die entry effects

A wide die gap, which reduces shear stress is the obvious solution to melt fracture (see **Figure 9.9**). However this leads to unbalanced orientation which compromises film properties. A number of other solutions can be considered.

The use of heated die lips (heat polishing) and polymer processing aids to reduce shear stress at walls is one suggestion. The easiest option is to blend LLDPE with LDPE. **Figure 9.11** shows the viscosity data at 230 °C for an octene LLDPE blended with general purpose LDPE in ratios of 40% and 60% LDPE. The apparent viscosity at 230 °C of the blends is significantly reduced as shown.

To test the effects of blending, a tubular section as shown in **Figure 9.12** based on a nominal 350 mm diameter die outlet section was used to compare the three blend ratios. In **Figure 9.13**, the shear stresses at the walls of this flow domain at 230 °C are calculated. The flow rate was 2.27 kg/hr/cm of die.

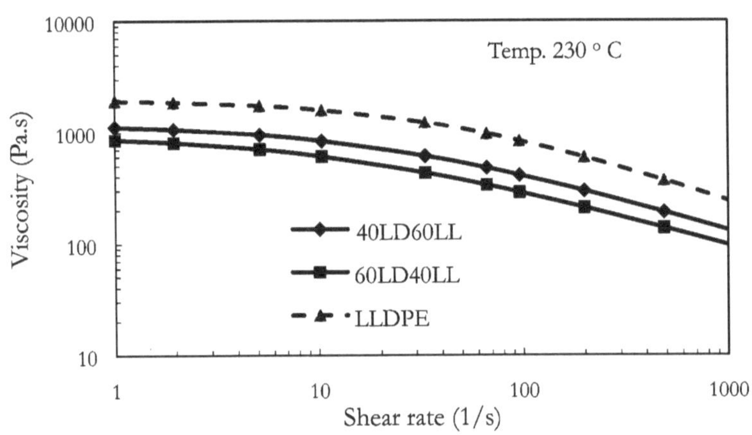

Figure 9.11

Shear viscosity of LD/LLDPE blends

The three blends will process satisfactorily with the 1.5 mm die gap. At 1.0 mm gap, only the linear lean (40%) blend may possibly run free of sharkskin. This data shows that by blending die gaps can be reduced well below 2 mm at commercial extrusion rates (2.27 kg/hr/cm).

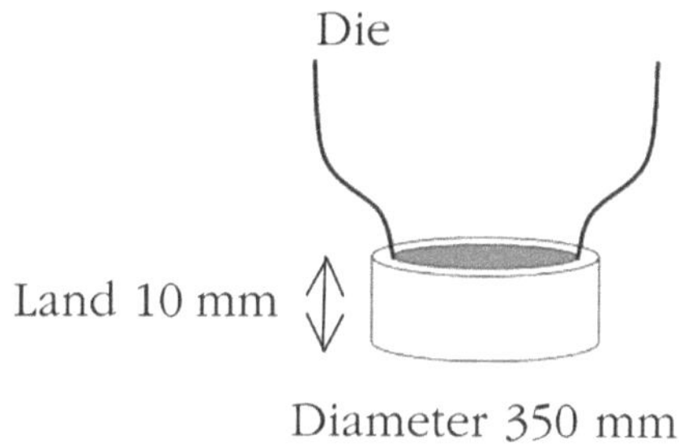

Figure 9.12

Simple exit die geometry

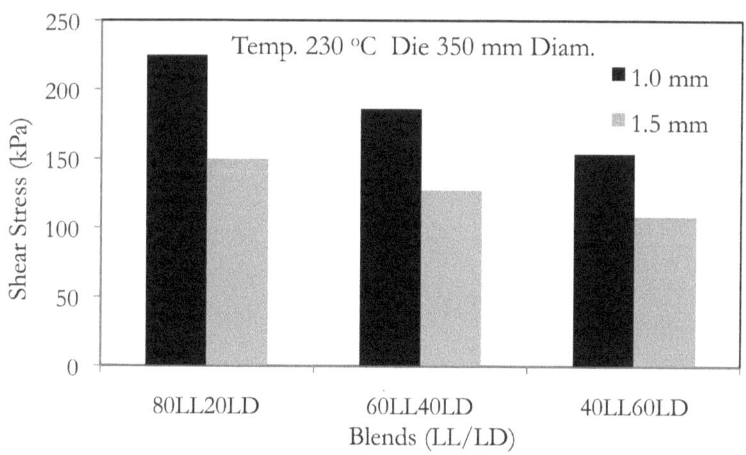

Figure 9.13

Shear stress at die walls

9.5 PPA and wall slip

The addition of polymer processing aids (PPA) to polyethylene to overcome sharkskin has been extensively researched over the years. The most successful are based on fluoroelastomers although other products such as boron nitride powder have been used. Fluoroelastomers are incompatible with polyethylene but have sufficient affinity to form stable two-phase systems. Small particles of the processing aid, a few microns in diameter, are dispersed in the LLDPE. As the polymer is extruded the processing aid contacts the metal surfaces and coats these with a constantly changing layer of fluoroelastomer. This layer reduces shear stress at the wall and consequently surface melt fracture is eliminated. Apart from the extra cost of the additive the experience with fluoroelastomers has not shown any disadvantages. Since the small particles of processing aid are embedded in the LLDPE matrix they do not change corona treatment response or sealing properties unless present in excess. Unlike slip additives, fluoroelastomers do not migrate to the surface to form a lubricating layer on the film. It is also claimed that die build-up can be reduced with PPA.

PPAs based on block copolymers of urethanes and polyols are also proposed as alternatives to the fluorinated polymers. These are lower in price and claimed to change boundary conditions at the die walls to generate slip and reduce shear stress. It is not clear how well these products compare to the fluoropolymer PPAs.

9.5.1 Evaluation of PPA using computer simulations

Under normal flow conditions the melt velocity at the wall is assumed to be zero. Wall slip will occur when a critical shear stress is reached, which causes the polymer at the wall to accelerate. The wall slip velocity is related to the wall shear stress by equation 9.1, which is valid for the isothermal case.

$$V = V_0 \left(\tau_w / 100\,\text{kPa} \right)^S \qquad (9.1)$$

τ_w Wall shear stress [kPa]
v_0 Reference slip velocity [mm/s]
s Constant [-]

By assuming a simple annular geometry (**Figure 9.12**) typical of the lip section of a blown film die, some estimates of shear stress with varying velocity at its walls were calculated. The results are summarized in **Table 9.2**. Calibration experiments should be used to estimate the values for the slip velocity and coefficient in equation 9.1.

At the mass flow rate conditions in **Table 9.2** the use of a 1 mm die gap will result in shear stresses at the walls of 296 kPa. Severe sharkskin is highly probable. If slip is allowed at the walls the velocity accelerates to 30 mm/s and the shear stress drops to 235/236 kPa. Sharkskin is still highly probable. By widening the gap to 1.5 mm the velocity at the walls is now 11 mm/s and the shear stress drops to ~173 kPa. At these conditions sharkskin looks less likely and can be eliminated by some adjustments to die temperature or further widening of the die gap. Without wall slip an LLDPE of the type used here at these conditions will require a die gap of 2 to 2.5 mm. (LLDPE (octene), MI = 1.0 dg/min., density = 0.919 g/cm^3).

This exercise demonstrates that if the PPA can cause the polymer melt to slip at the walls reducing the die gap is possible.

Table 9.2

Effect of slip on wall stress

OD Diameter (mm)	352	352	353
ID Diameter (mm)	350	350	350
Land (mm)	10	10	10
Die gap (mm)	1.0	1.0	1.5
Mass flow rate (kg/hr)	250	250	250
Temp. (°C)	220	220	220
Results	**No Slip**	**Slip**	**Slip**
Shear stress @ walls (kPa) in/out	296/296	235/236	173/172
ΔP (MPa)	5.9	4.7	2.3
Velocity @ walls (mm/s)	0	30	11

In **Figure 9.14** the shear stresses with slip are compared at four die gaps. The octene LLDPE in **Table 9.2** is entered. The calculation shows that a die gap of 1.8-2.0 mm should be sufficient to avoid sharkskin formation with this LLDPE grade. At 1.5 mm the condition is marginal.

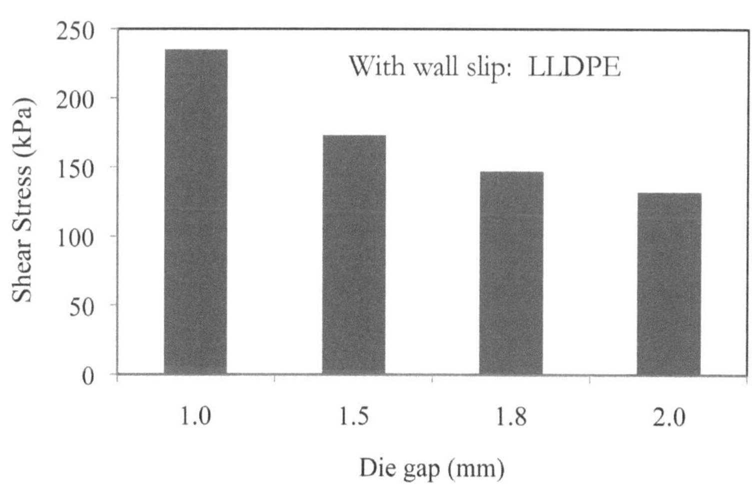

Figure 9.14

Effect of die gap on shear stress

In **Figure 9.15** two metallocene LLDPEs are compared to two Z-N LLDPEs. All are characterized by a melt index of 1.0 dg/min. and density of ~0.918 g/cm³. The die gap is fixed at 1.5 mm and the entries are as in **Table 9.2**. With slip the shear stress is reduced as shown. These virtual experiments predict that sharkskin will not be eliminated with slip at 1.5 mm die gap. The hexene mLLDPE develops the highest shear stress in this group.

In **Figure 9.16** two wider die gaps are evaluated with slip at the walls. The results show that the two octene resins may be able to avoid sharkskin with the 1.8 mm die gap. The hexene grade will almost certainly generate sharkskin. A die gap of 2.0-2.5 mm would be recommended with the butene and hexene resins. Further reductions in shear stress are possible by reducing throughput and increasing temperature. Both of these options are undesirable.

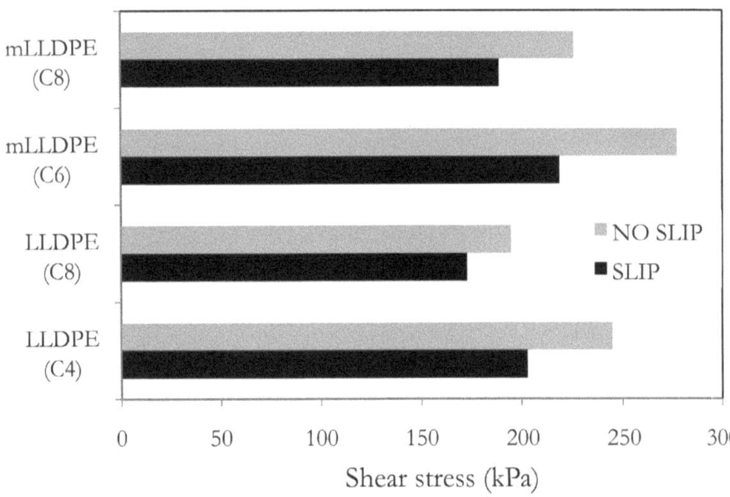

Figure 9.15

Effect of slip on shear stress for various PEs (1.5 mm die gap)

These virtual experiments indicate that as the addition of PPA reduces shear stress through slip at the walls narrower die gaps can be used. With LLDPEs, die gaps above 1.5 mm will still be required at the throughputs used in these simulations. From this data the shear viscosity of commercial LLDPEs, Z-N or metallocenes, can vary considerably.

(In these simulations each material is characterized principally by its shear viscosity. The Carreau correction was applied using the raw rheological data in all cases. However, the data was obtained from different sources and may lack consistency in comparing these four polymers.)

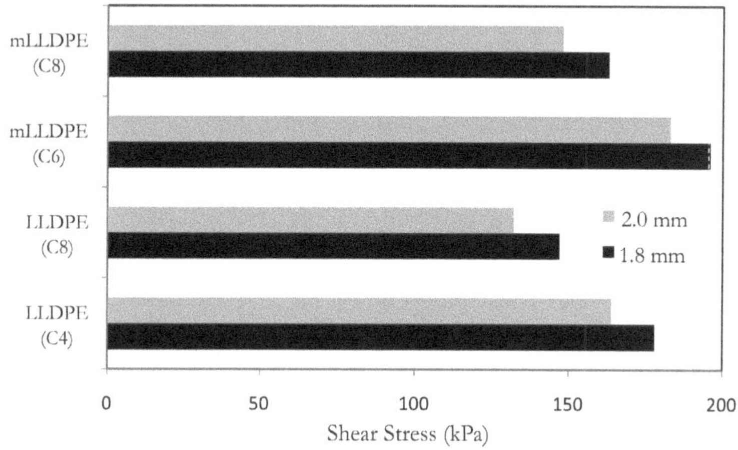

Figure 9.16

Shear stress with slip at 2 different die gaps (350 mm diameter die, flow rate: 250 kg/hr)

For the processor it is also important to remember that the use of melt index alone may be an insufficient guide to sharkskin without access to the shear viscosity data of the polymer. In borderline cases even if an LLDPE of a certain melt index does not form sharkskin another grade with the same melt index may develop sharkskin at the same processing conditions.

The recommended concentration of PPA is 300-500 ppm. These are available as concentrates containing 3-5% PPA.

In **Table 9.3** some surface tension measurements of LLDPE films with and without PPA addition are shown. The data shows that the presence of an excessively high

Table 9.3

Effect of 1500ppm PPA on corona treatment

Sample	Die gap (mm)	Surface tension (mN/m)
Control	2.0	48
PPA-1	2.0	48
PPA-2	2.0	46
Control	0.6	48
PPA-1	0.6	48
PPA-2	0.6	48
PPAs (fluoroelastomers) are best blended with LLDPE-based masterbatches		

amount (1500 ppm) of PPA does not have any detrimental effect on surface treatment within the error of the test method. Elemental analysis of the surfaces by ESCA showed the absence of fluorine confirming that the PPA does not migrate to the surface of the film. Film physical properties are not adversely affected by the presence of these low concentrations of fluoropolymers.

Finally, it has been reported that catalyst residues and acid scavengers such as hydrotalcite used in the manufacture of LLDPE can adversely affect the performance of the PPA. This might explain some of the reported variations in the use of these products.

There are patents that claim that further improvements in the performance of the fluoroelastomer can be achieved in combination with boron nitride powder. Mixing boron nitride with the fluoroelastomer is believed to have a synergistic effect in retarding the formation of melt fracture.

The benefits of die surface treatments to reduce COF and minimize shear stress at the walls are under continuous evaluation within the industry. Titanium nitride coatings are claimed to significantly reduce shear stress at the die walls. A 15% reduction is claimed over 4140 steel. The latest developments claim that boron nitride treatment can reduce the surface COF to an even lower level.

9.6 Rotating or fixed dies

The oscillating or rotating die is the most widely used method to spread out gauge variation round the circumference of the bubble to yield a flat film roll of uniform hardness across its width. It is important to remember that rotation does not eliminate thickness variations, but simply randomly redistributes them along the length of the film roll thus avoiding the build-up of localized gauge bands. Oscillation through a full 360° is usual. The advantages of a rotating die over a fixed die with rotating haul-off are:

- The winder is placed at the same level as the extruder to facilitate operator control.
- Less friction is developed between the bubble and collapsing frame. Rotating the collapsing frame/nip roll section introduces more friction on the film's surface, which can be problematic with films of high coefficient of friction (COF).

By taking precautions such as constructing a draft excluder enclosure round the bubble and carefully adjusting the die lips and cooling ring, good results are obtained with the rotating die. There are, however, some disadvantages to this layout:

- The rotating/oscillating die will not redistribute temperature variations or changing velocities in the melt stream.
- Gauge variations caused by the ambient surroundings, e.g. air turbulence, of the machine are not redistributed.

Furthermore, when using rotating multilayer coextrusion tubular dies more disadvantages have been identified:

- The die height is increased by around 1 m, which increases melt residence times, resulting in degradation of heat-sensitive polymers.
- An increased proneness to leaks at the seals of the rotating mechanism.
- Stripping and re-assembly for cleaning and maintenance is time consuming.
- Purge times can be excessive.

As a result of these observations there has been renewed interest in the use of fixed dies with rotating haul-off. Improved designs in turning bar systems to deliver the layflat to the winder placed at the same level as the extruder are now offered by all major machinery suppliers.

Fully automated coextrusion blown film lines based on fixed dies are becoming quite common. These have rotating collapsing frames and nip roll units, which are equipped with turning bar systems that guide the layflat down to the winder placed adjacent to the extruder(s) thus eliminating the principal disadvantage of fixed dies. Full 360° oscillation is claimed. Friction is minimized by using rotating brushes and other proprietary design features in the collapsing frame.

Some manufacturers are offering fully automated lines with internal bubble cooling (IBC) for up to 7-layer structures.

9.7 Automatic gauge control

Although oscillating the die or haul-off will redistribute the thickness variations in the layflat to yield flat rolls, problems in converting and tension variations leading to wrinkles can still be caused by thickness fluctuations. More significantly, the nominal gauge has to be increased to the minimum performance specifications thus unnecessarily increasing raw material usage. This has led to the search for accurate automatic gauge control systems.

An automatic gauge control system must control the following three elements, which typically contribute the following proportions of thickness variations in blown film.

Die lips	1 to ~10% variation
Die port lines	5 to ~15% variation
Ambient air	5 to ~12% variation

Several systems for automatic gauge control of the blown film have been devised with varying degrees of success. Tthe best systems can deliver reduced average gauge deviations of ~50% compared to manual adjustments.

The advantages of automatic film thickness control are:

- Different flow properties of the polymers are compensated
- Film flatness is improved
- Curvature is reduced
- Thickness tolerances in TD can be reduced to the level of MD tolerances

The gauge control systems are of two basic types:

1. Die-based systems using cartridge heaters round the circumference just below the lips. These are activated by thickness readings determined by a rotating gauge. The melt is heated in specific zones allowing high spots to draw down further.
2. Air ring systems, which adjust localized air flow to correct thickness variations via feedback from a rotating gauge. These are the most popular because of their lower cost. The best designs are capable of thickness control to within ±5% from the average value.

Chapter 10

Tubular film layout

Figure 10.1 illustrates the basic machine layout for the tubular/blown film process. Melt temperature with PEs is 220-240 °C. The bubble is collapsed into a two layer layflat film, which is easily converted into bags and sacks. In **Figure 10.2**, the film is slit into single layers for applications such as lamination.

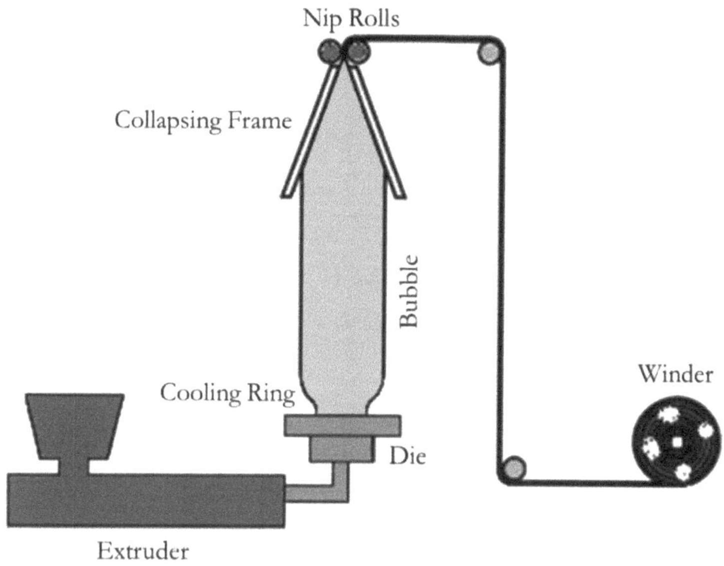

Figure 10.1

Schematic of a blown film line

The extruder delivers the melt into the die, which forms it into a tube. The tubular film is then nipped as it exits the die. Cold air is injected through an orifice in the centre of the die body to inflate the tube into a bubble, which is allowed to stabilize. The bubble is cooled rapidly with cold air from a cooling ring. A pair of driven nip rolls placed a few meters above the die pulls the melt and draws it down to the required thickness. It may be necessary to use water cooled nip rolls. The final film thickness is controlled by the rotating speed of the nip rolls, the extruder output and by the inflation of the tube. Extruder output is set by its screw speed. The collapsing frame assists to feed the bubble into the

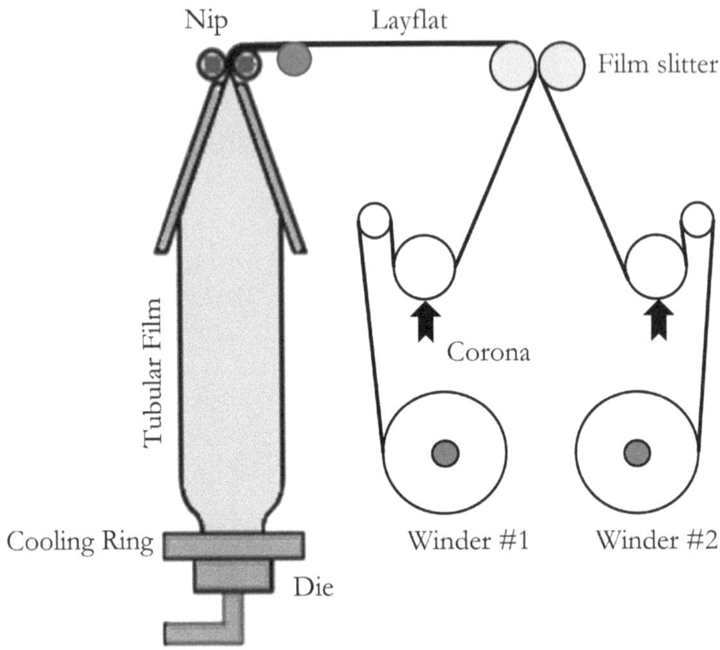

Figure 10.2

Layout for single-layer film

nip rolls where it is pressed into a layflat (2 layers). The space between the die, the collapsing frame and the nip rolls is termed the tower. It is essential that the centre of the nip rolls be aligned with a plumb line dropped over the centre of the die. Otherwise film wrinkling will occur.

The blow-up ratio (BUR) is the ratio of the circumference of the die to the circumference of the bubble and is calculated as in equation 10.1.

$$BUR = \frac{2 \times L}{\pi \times D} \qquad (10.1)$$

L = Layflat width D = Die diameter

The circumference of the film is in effect 2× the layflat width.

The range of blow-up ratios used with LDPE and LLDPE are typically 1.8 to 2.2:1 and controlled by the volume of air pumped into the bubble. Higher blow-up ratios may be used for shrink films. With HDPE blow-up ratios are 3.5-4:1. The nip rolls must completely seal off the air in order to keep the bubble at a constant internal pressure. Dual lip cooling rings and chilled air would be recommended with all the linear PEs to optimize stability. A corona treater if required will be placed at the most convenient point between the nip rolls and the winder. The film can be one side or two side treated. It is important that there is no air entrapment between the layers otherwise back treatment can occur. The slitter knives are placed before the second pair of driven rolls, which controls the web tension prior to feeding to the winders.

The requirements for high quality PE films are as follows:

- Flatness (low gauge variation)
- No film curvature
- Non-stretch (stiffness)
- Corona treatment to accept inks and adhesives
- Precisely controlled slip and antiblock formulation
- Heat-sealing: important for packaging and lamination applications
- Low seal initiation temperature (SIT)
- Wide sealing range
- High hot tack strength
- No contamination
- No static ($< 10^{11}$ ohm surface resistivity): important for packaging powders
- Transparency/low haze: optional
- No gels, unmelts, etc.: critical for low gauge films

The films must be corona treated to a surface tension of 42-45 mN/m in order to interact effectively with printing inks or adhesives used in the lamination process. Slip and antiblock agents are added to improve machineability and openability of the layflat film. The PE manufacturers will supply their film grades ready formulated. These are usually described as low, medium and high slip. There are many variations depending on the application and the type of polyethylene. Some processors will formulate their polymers with specialized masterbatches.

For best results thickness variation should be kept below ±10% of the target value. For some applications such as lamination film it may be necessary to reduce the variation to ±5%. These tight gauge requirements can only be met with precision designed dies and cooling rings plus automatic thickness control systems.

To avoid gauge bands on the finished rolls rotating or oscillating dies are used. A typical rotating die cycle is 7-8 minutes. Fixed dies with rotating haul-offs will have 360° oscillation and are equipped with turning bar systems that guide the layflat down to the winder placed adjacent to the extruders. The rotating time for the full cycle is around 20 minutes.

The collapsing and winding systems must deliver flat rolls wound at uniform tension. Centre or gap winders are preferred. Coextrusion facilities can be added to save resin cost and to enhance the performance of the final film.

It is essential that the extruders and cooling tower be fixed on strong foundations. The foundations must be sufficiently rigid to avoid any movement which can cause misalignment between the die, the collapsing frame and nip rolls during operation. Wrinkles and poor gauge control can occur from small movements of the collapsing frame and nip rolls relative to the die.

10.1 Bubble stability

Frost line heights for LD/LLDPE are typically 2-3 D and 8-12 D for HDPE (long neck or wine glass process). The frost line height must be set to achieve the best bubble stability as the melt flows from the die exit to the freeze line. The Venturi effect generated by high velocity air flowing from the cooling ring creates a low pressure zone, which intensifies air flow in the area between the die face and the orifice. This effect improves cooling efficiency and is one of the key principles guiding the design of cooling rings. **Figure 10.3** shows a typical bubble cooling regimen for an LDPE extrusion. In this example, the effect of the cooling air is to drop the melt temperature at the die exit from 210 °C to 160 °C. At the frost

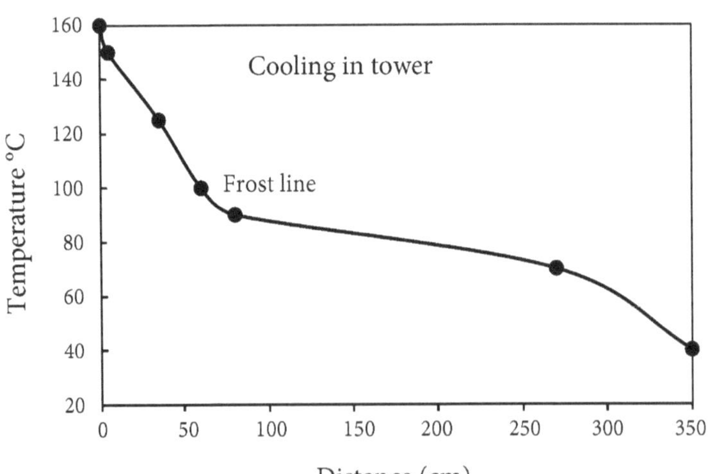

Figure 10.3

Bubble cooling from the die to the nip rolls

line the temperature is ~100 °C. By the time the film reaches the nip rolls its temperature is down to 40 °C.

A computer generated bubble profile is plotted in **Figure 10.4**. The process conditions are in **Table 10.1**. In the following simulations no account is taken of changes in elongational viscosity, which occur from the die exit to the frost line.

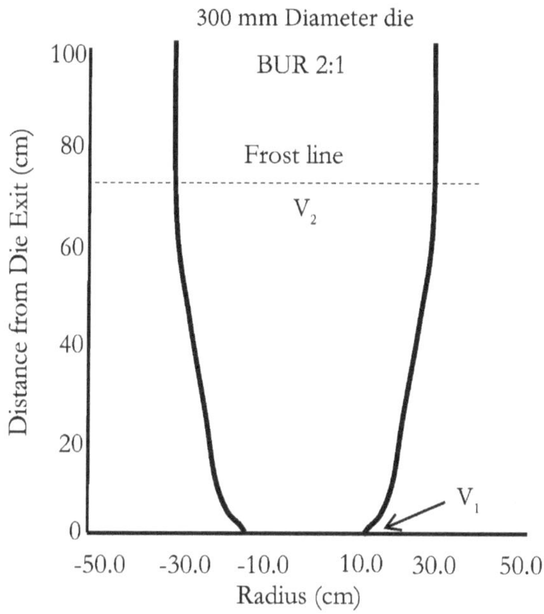

Figure 10.4

LDPE Blown film bubble

Table 10.1

Film process conditions

LDPE mass flow rate (kg/hr)	200
Temperature (°C)	200
Die diameter (mm)	300
Die gap (mm)	1.0
Film thickness (μm)	30
Blow-up ratio	2:1
Frost line (mm)	750
Line speed (V_2) (mpm)	65
Velocity @ lips (V_1) (mpm)	4.65
Cooling ring air temp. (°C)	20

A 30 μm LDPE film is produced at 200 kg/hr at a line speed of 65 mpm. The frost line height is 750 mm from the die face. The velocity at the lips is calculated from the die diameter (W) and gap (t) and mass flow rate (Q) as in equation 10.2. Melt density is assumed to be 760 kg/m^3.

$$Q = V \times t \times W \times \rho \qquad (10.2)$$

Velocity = V (m/s) Die gap = t (m) Die width = W (m)
Output = Q (kg/s) Melt density = ρ (kg/m^3)

The stretch ratios at 1.0 and 2.0 mm die gaps are calculated and shown in **Table 10.2**. The total stretch ratio (TS) is the product of the MD and TD ratios (equation 10.3).

$$TS = MD \times TD \qquad (10.3)$$

TS is calculated from V_2/V_1 and the TD is the BUR.

Table 10.2

Stretch ratios at 2 die gaps

Die gap (mm)	**1.00**	**2.00**
V_1 mpm (velocity @ die lips)	4.65	2.33
V_2 mpm (line speed)	65.00	65.00
BUR (TD stretch)	2.00	2.00
Total stretch (V_2/V_1)	13.98	27.96
MD stretch	**6.99**	**13.98**

The above calculation shows that increasing the die gap to 2 mm increases the MD stretch ratio from 7 to 14 at 2:1 BUR. This will have a negative impact on properties and thickness distribution. This is one of the disadvantages of LLDPE film extrusion which requires wide die gaps to avoid sharkskin formation.

Figure 10.5 shows computer generated plots of shear stress and strain rate developed from the die exit to just above the 75 cm frost line. The stress rises as it approaches the frost line and then plateaus as the polymer solidifies. Similarly, the strain rate rises to a peak and decreases rapidly as the melt begins to solidify. The film can no longer be deformed. An ideal melt should develop low stress at the strain rate peak and high stress as the film reaches the freeze line and crystallizes. If excessive stress develops at maximum strain melt rupture will occur. It is in the zone below the frost line as the stress rises rapidly that polymer chain uncoiling occurs and determines the degree of orientation and consequently

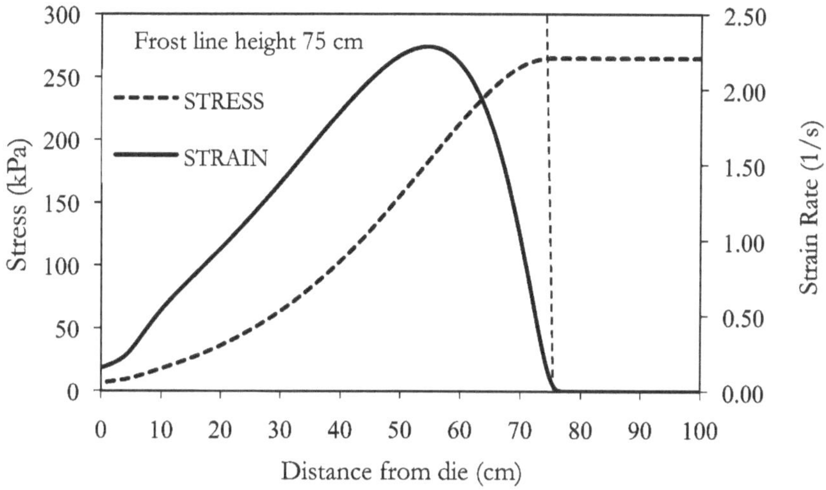

Figure 10.5

Stress/strain curves from the die exit to the frost line

film properties. At the die exit, where the temperature is highest chain slippage occurs with very little film orientation.

LDPE is strain hardening and there is, therefore, the possibility that as the strain increases rapidly a destructively high stress develops causing the melt to break. LLDPE (linear) in contrast is tension thinning and stress will decrease with increasing strain rate. This explains why at equivalent melt index LLDPE resins will have better drawdown than LDPE, i.e. the ability to produce thinner films. In practice, however, most melt breaks occur in the presence of gels, unmelts and hard particles, which act as localized stress raisers leading to rupture. Polymer cleanliness is a prerequisite for efficient operation.

Blending LDPE with LLDPE is one way of optimizing melt strength and drawdown. This balances the strain hardening effect in LDPE with the strain thinning element in LLDPE. A method to improve the bubble stability of linear polymers is to synthesize them with wide or preferably bimodal MWD. Bimodal structures will improve bubble stability as the HMW element will ensure that a high viscosity is reached at the freeze line and the lower MW component will avoid excessive stress at the maximum strain rate reached before the film solidifies. This approach has been very successful with high molecular weight (HMW) HDPE.

10.2 HDPE blown film

HDPE films including HMW grades will tear easily particularly when formed into thin films. Orientation when applied at the correct conditions will induce a very significant degree of reinforcement into the film. Biaxial orientation will improve tensile strength, puncture and impact resistance. Therefore by the combination of high molecular weight (HMW) and a high degree of biaxial orientation very strong low gauge films can be extruded from this linear polymer. State-of-the-art lines capable of delivering 375 kg/hr of HMW-HDPE from 90 mm diameter extruders at line speeds of 200 mpm are available.

Figure 10.6 shows a computer generated HDPE blown film bubble extruded from a 120 mm diameter die. The mass flow rate is 200 kg/hr at a melt temperature of 230 °C. The long neck is maintained at a frost line height of 120 cm (10 D). In the long neck section the orientation is principally in the MD and is dependent on the velocity ratio of V_1 to V_2. The cross direction or TD orientation is induced in the zone between V_2 and V_3. The blow-up ratio is 4: 1. The film is drawn from a 1.2 mm die gap to 20 μm film.

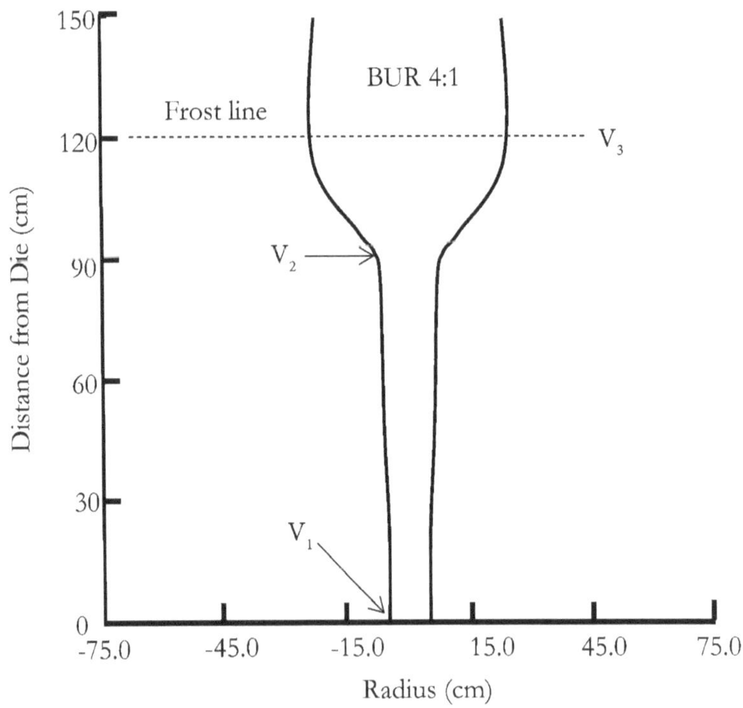

Figure 10.6

Long neck HDPE film extrusion

Figure 10.7 shows the bubble cooling regimen from the die face to just above the frost line. The temperature drops catastrophically in the first 20 cm from the die, caused by the cooling air and begins to level off followed by an increase in cooling rate as the bubble is expanded and approaches the frost line. Approaching the frost line the film temperature is 131 °C and rapidly crystallizing. At this point the film thickness is 20 μm travelling at a web speed of 142 mpm.

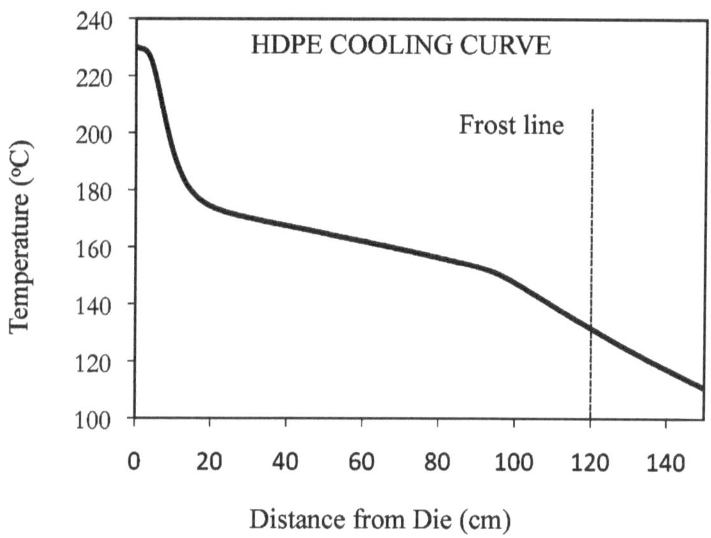

Figure 10.7

Bubble cooling curve for long neck process

Figure 10.8 shows the MD and TD strain rate developed from the die to the frost line. The strain rate in the MD rises from the die and then dips at around 90 cm distance where the tube expands. In the TD, the strain rate is minimal at the same distance from the die face and then peaks rapidly as the MD drops. In this region the bubble is expanded to a blow-up ratio of 4:1. Both curves peak before the frost line (120 cm) and drop to zero as the melt solidifies. The critical orientation area is between V_2 and V_3 where the film temperature is approaching the crystallization point and the applied stress is reaching its maximum. The TD orientation in the final step is induced in this region. However, as **Figure 10.8** shows orientation is also induced in the MD. The minor difference between the peak heights indicates that the film is slightly more oriented in the TD. **Table 10.3** shows some of the computerized data. The film accelerates from 9.4 mpm at the die to 142 mpm at the frost line.

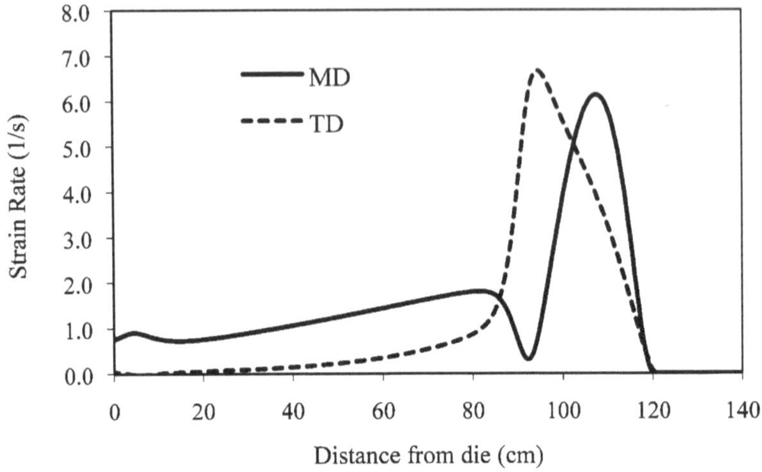

Figure 10.8

MD/TD Strain rate developed in bubble

Table 10.3

Velocity and temperature profile in long neck process

Velocity	Position	Velocity (cm/s)	Velocity (mpm)	Distance (cm)	Temp. (°C)
V_1	Die lips	15.7	9.4	0	230
V_2	Inflation	122	73	90	153
V_3	Frost line	236	142	120	131

The stretch ratios in the three zones of the process (**Figure 10.6**) are as follows:

V_3/V_1 = 15 (total stretch ratio)

V_2/V_1 = 7.8 (stretch in neck section)

V_3/V_2 = 1.9 (stretch in expanded section)

The total stretch (TS) calculation was given in equation 10.3.

$$TS = MD \times TD \quad (10.3)$$

The total stretch ratio (TS) is 15 from V_3/V_1. From a BUR of 4:1, the MD stretch ratio is 3.75 (TS/TD). The film will be perfectly balanced in both directions when MD/TD is unity. The ratio of MD to TD is calculated from equation 10.4.

$$MD / TD = 3.75 / 4.0 = 0.94 \quad (10.4)$$

This result indicates that at a BUR of 4:1, the film is slightly more oriented in the TD. If the BUR is reduced to 3.75, the film will be perfectly balanced. The differences between the two peak strain rates in **Figure 10.8** confirm this outcome. (Side effects from die swell as the tube exits the die are not taken into account.)

Summing up this data leads to the following conclusions:

- BUR should be set in the range of 3.5-4:1. Higher BUR will lead to excessive TD orientation.
- The velocity V_2 depends on the height of the neck at any given condition, i.e. acceleration from the die lips.
- If the neck height is too low, V_2 will be low, which will decrease the MD stretch in this zone. The neck height will also control the film temperature as it is expands in the zone just below the frost line.
- **Figure 10.7** shows the film temperature at ~150 °C as the tube is expanded 90 cm from the die face. The melt is now close to its crystallization temperature. The temperature up to the frost line is critical in inducing the stretching forces on the bubble to achieve maximum uncoiling of the polymer chains.

Frost line heights of 8-12 D are usually recommended in the literature.

10.3 Bubble cooling systems

Modern blown film lines have air cooling inside and outside of the bubble. The position of the rings and their adjustments will regulate the form of the bubble. The cooling equipment has the following objectives:

- Solidify the film
- Maintain minimum gauge variations
- Maximize production rate

The cooling system irrespective of its design must be able to cool the melt from 220-240 °C at the die exit to 40 °C or less as the film enters the nip rolls.

10.3.1 Effect of air temperature on output

The effect of lowering cooling air temperature from 20° to 10 °C was tested with PC simulation software. The die exit parameters are in **Table 10.4**.

Table 10.4 shows that at 200 kg/hr of LDPE output the film cools to 40 °C at a height of 258 cm from the die face with an air temperature of 20 °C. If the cooling air is chilled to 10 °C, output is increased to 245 kg/hr with the film cooling to 40 °C, at very nearly the same distance from the die. This simulation shows an output increase of 22.5% illustrating the benefits of using refrigerated air.

Table 10.4

Effect of air temperature on output

Cooling ring air temp.	**20 °C**	**10 °C**
Film thickness (μm)	75	75
Melt temperature (°C)	220	220
Frost line (mm)	750	750
Line speed (mpm)	31	38
Film height @ 40°C (cm)	258	255
LDPE mass flow rate (kg/hr)	200	245

General troubleshooting rules regarding the frost line are outlined below.

The frost line should always be level and even, and a tilted frost line is due to:

- Die not centred under the lips
- Air ring not level
- Worn die body
- Restricted ports

A jagged frost line (**Figure 10.9**) is due to:

- Worn die/port flow effects
- Restricted die ports
- Dirty air ring
- Poor plenum design affecting air flow
- Uneven plenum hose lengths

Figure 10.10 shows schematically the three types of air cooling ring designs commonly in use with PE resins. Type A is used with LDPE and will yield a reduced output of 1.5 kg/hr/cm with LLDPE. This design will lead to bubble instability problems with low melt strength LLDPE. Type B with dual orifice is recommended for LLDPE and will yield an output of 2.2 kg/hr/cm of die. The use of the air collar, which acts as a secondary cooling ring in C (stacked air rings),

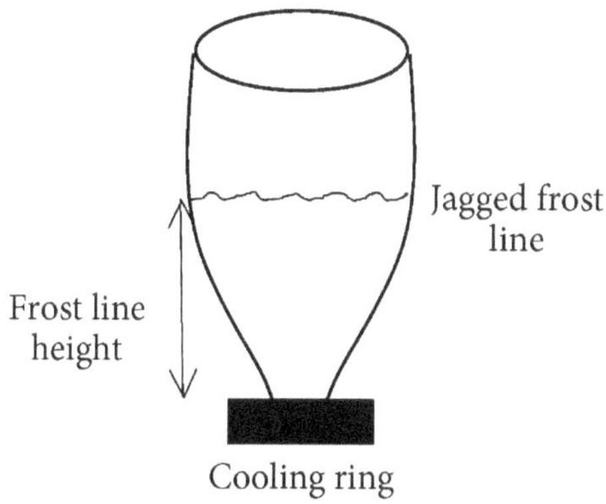

Figure 10.9

Jagged frost line

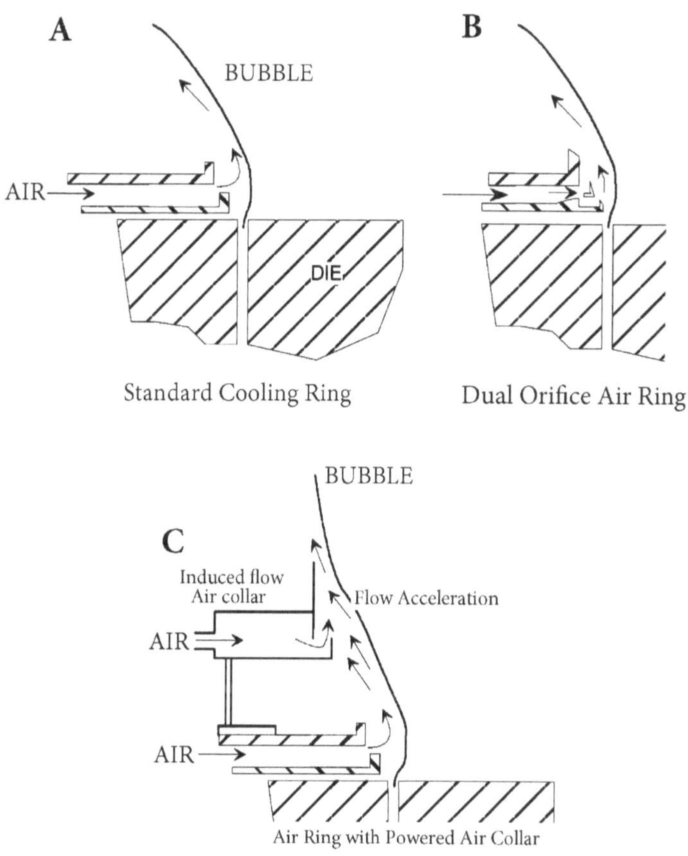

Figure 10.10

Cooling ring design

is recommended for heavy gauge films. This system will achieve an output of around 2.0 kg/hr/cm with LLDPE.

In **Figure 10.11** a cross section of a typical single lip cooling ring with adjustable iris diaphragm is shown. The adjustable diaphragm can be fitted with the three basic designs above and will assist in adjusting the shape of the bubble and may also improve stability.

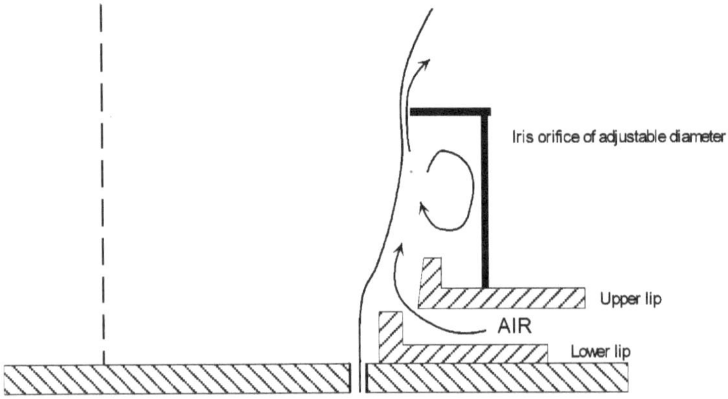

Figure 10.11

Heat transfer with iris diaphragm

To further improve cooling efficiency internal bubble cooling (IBC) is added to the system. The basic layout is shown in **Figure 10.12**.

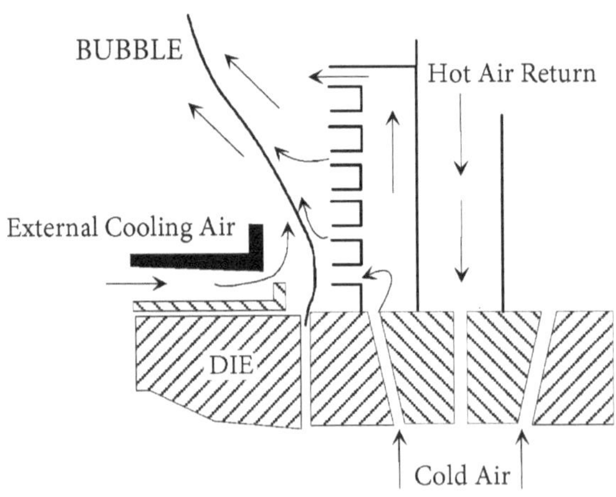

Figure 10.12

Internal bubble cooling

The principle underlying internal bubble cooling (IBC) is to apply a flow of cold air both inside and outside of the bubble. The bubble is more efficiently cooled and stabilized by the opposing air currents. Leading manufacturers of IBC systems claim an increase in output of more than 50%. The bubble air pressure must be kept low. Melt fracture will occur if the die lips are chilled. Several systems are available from different suppliers which claim certain advantages depending on the type of polymer and film produced. When used with LLDPE the IBC units is claimed to achieve outputs of 3-4 kg/hr/cm of die compared to 2.3 kg/hr/cm with a conventional dual orifice cooling ring.

Using a computerized heat exchange simulation the effectiveness of an IBC was compared to a conventional cooling ring. Cooling air temperature was set at 20 °C in both cases. The results are in **Table 10.5**. With a standard cooling ring, the LDPE film temperature at 1 m distance from the die face is 98 °C. With the IBC programmed into the simulation throughput was increased from 200 kg/hr to 250 kg/hr with the film temperature held at 98°C at the same distance from the die exit. This result represents a 25% increase in output. This suggests that claims of increases of 50% may be optimistic unless refrigerated air is used with the IBC.

Table 10.5
Effect of IBC on outputs

Conditions	Normal	IBC
Melt temperature (°C)	200	200
Film thickness (μm)	30	30
Frost line (mm)	750	750
Film temperature @ 1 m (°C)	98	98
Cooling ring air temperature (°C)	20	20 (in/out)
Line speed (mpm)	65	82
LDPE mass flow rate (kg/hr)	200	250

The IBC system must perform the following operations:

1. Since the internal bubble air is continually exchanged, the bubble diameter must be automatically controlled with built-in sensors, usually ultrasonic. With the best designs currently available the layflat width can be controlled to within ±0.15%.
2. Blowers are provided to:
    - Introduce cold air inside the bubble

- Remove hot air from the bubble (exhaust blower)
- Provide cooling air to the external cooling ring

3. The layflat width or bubble diameter is controlled accurately by the exhaust blower.

Cold air supply and the air exhaust tubes must be well insulated from the die body. Finally low melt strength polymers (LLDPE) tend to be more sensitive to the IBC settings.

10.4 Collapsing frame

To avoid creases and wrinkles forming in the film it is important that the collapsing frame design is selected on the basis of the film stiffness and the coefficient of friction (COF) of the outside bubble surface. For medium stiffness films, e.g. LDPE and LLDPE, wood slats are effective and cost efficient. As COF increases (EVA, POE) various designs of freely rotating roller surfaces have proved effective. However, bare metal rollers should be avoided as they can cause non-uniform heat transfer across the film. Hard plastic frictionless roller systems are the most effective. When very rigid polymers (modulus >1400 MPa) and thick films are run, the stiffness in the bubble creates the need for a four sided system with a square-off section to make all collapsing lengths equal. This will greatly reduce the tendency to form wrinkles and creases.

Collapsing tubular film into a layflat is much easier with LDPE than with HDPE. The high stiffness of HDPE requires high stresses to deform the tube into a layflat, which results in irreversible strain in the film. The inability to reverse the strain results in the following defects with HDPE and other stiff materials:

- Creases or wrinkles in both the MD and TD
- Sagging of the layflat
- Slack edges

As the bubble is transported towards the nip rolls the film is gradually deformed from a circular (tube) to an elliptical shape and finally a flat virtually two dimensional section. In the process the film is subjected to friction. The area of friction is increased as the film approaches the nip rolls and is flattened as shown in **Figure 10.13**.

Sagging and slack edges are caused by this imposed deformation on the tube as shown. The sketch shows the strain build-up at the circumference of the bubble until it is finally flattened. The strain is higher at the edges and the centre.

Reducing the collapsing angle reduces the strain as shown in **Figure 10.14**. However, reducing the angle will also increase the surface area of the frame in contact with the film. If the film has a high COF drag will increase in the frame and will distort the film flatness resulting in baggy rolls at the centre. Ideally the surface should have a zero coefficient of friction.

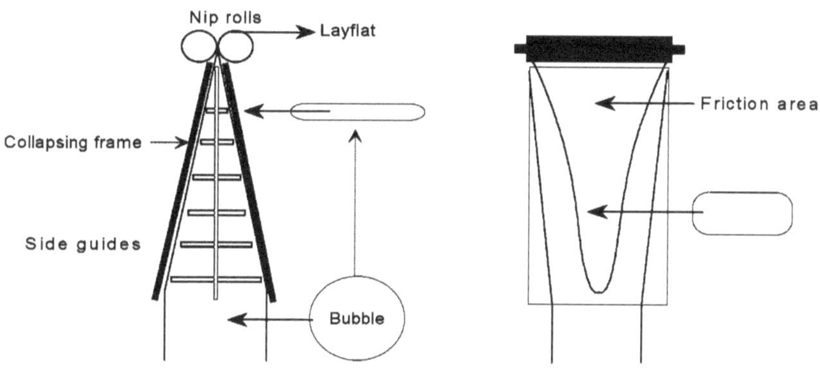

Figure 10.13

Collapsing process

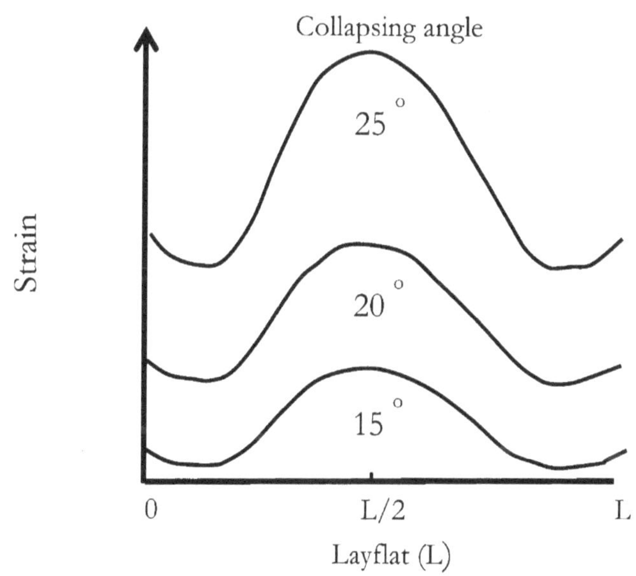

Figure 10.14

Strain across layflat width

Since the strain on the bubble is not completely elastic it is, therefore, transmitted as stress back to the area around the frost line of the film. The higher the E-modulus of the film the greater the stress at the freeze line, and the plastic deformation becomes more pronounced. This permanent deformation results in the collapsing defects listed above. Keeping the film hot as it feeds into the nip reduces its stiffness, which reduces the stresses as the film is collapsed. Therefore with HDPE short tower heights are used to minimize the heat loss of the bubble.

As coextrusion blown film processes increase in complexity with combinations of polyamides, EVOH and low melt strength polymers, collapsing frames employing carbon fibre-based rollers are claimed to be in increasing demand. With soft films such as EVA and very low density mPEs longer collapsing frames with a collapse angle of about 7° from the vertical are generally required as internal bubble pressure and melt strength are lower. The effects of heat transfer (loss) on the bubble are significantly increased. Multilayer structures incorporating polyamide are extruded at elevated temperatures also requiring longer collapsing frames. Under these conditions it is better to reduce the rate of heat transfer from the bubble to minimize wrinkling and thickness variations.

The heat transfer rate of carbon fibre rollers is approximately 10^3 times lower than that of aluminium rollers. This much lower conductivity draws heat away from the film much more slowly and uniformly than metal rollers. Since the film first comes in contact with the collapsing frame and solidifies, wrinkling will be minimized by the slower and more uniform cooling rate. Another advantage of the lighter carbon fibre rollers is their lower inertia, which also helps to reduce wrinkling.

10.5 Edge fold strength

The edge folds of blown PE films almost invariably present inferior mechanical properties to the film area between them. This applies in particular to dart impact and tear strength. As a generality impact strength at the crease averages around 80% of the overall film toughness.

The processing parameters that affect edge fold strength are as follows:

Frost line distance and cooling rate. The higher the frost line or conversely the slower the cooling rate the higher the crystallinity and stiffness and consequently the weaker the crease.

The more the film is oriented in a given direction, the weaker the fold strength. This infers that low crease strength in the MD, which is the most commonly occurring problem can be improved by:

- Increasing BUR
- Reducing line speed
- Maintaining a bubble shape as shown here:

rather than:

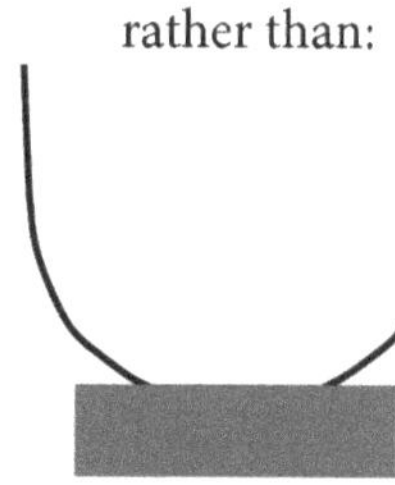

Recommended temperature range at the nip rolls: 28-38 °C. Too low a temperature leads to brittleness, whilst excessively high temperatures result in a sharp edge fold caused by applying the nip pressure on a soft warm film. Normally, edges should be round shaped to minimize splittiness. The rolls will normally be water cooled.

The higher the nip pressure, the weaker the edge fold. The nip roll pressure should be just sufficient to prevent air escape. For a given force applied to the roll, softer rubber rolls give a greater contact area and hence less pressure. This benefits fold strength. The larger contact area provided by a softer rubber roll requires less pressure to prevent air from escaping.

Nip Roll Options

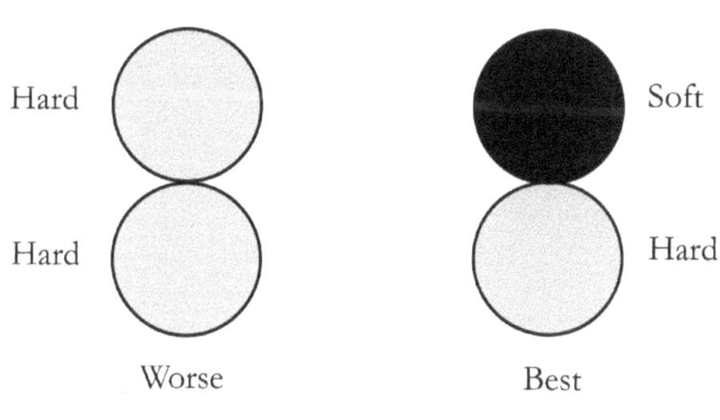

As a guideline nip pressure should be held at 1.5-2 bar. For thick heavy duty film (>150 μm) it is recommended to maintain a slight free space between the nip rolls corresponding to a fraction of the film thickness, so as to prevent excessive squeezing of the film.

10.6 Winding tension

Winding tension should be less than 0.5% of the film's 1% tensile secant modulus. **Table 10.6** shows the relationship between the modulus of four PE products and the recommended winding tension for 1 m wide 50 μm thickness film.

Table 10.6

Winding tension of four PE films

Films	Sec. modulus (g/mm^2)	Max. tension (g/mm^2)	Max. tension (μm g/cm × μm)	Permitted max. load on 1 m 50 mm film (kg)
LDPE	18000	90	0.9	4.5
LLDPE	24000	120	1.2	8
HDPE	52000	280	2.8	13
LD/LL/HD 33%/34%/33%	32000	180	1.8	8

The above data shows the desired winding tensions for the single-layer 50 μm thick film arriving at the winding station. However, the tension for the layflat (2 layers) between the two pairs of nip rolls should be calculated on the basis of twice the figures in Table 10.6, since the total thickness is now 100 μm. A further complication is that in the upstream part of the machine the film is some 10-20 °C higher in temperature. The tension should, therefore, ideally be some 80-90% lower than the calculated value.

Table 10.7 summarizes the effects of the extrusion process on some critical film defects.

Table 10.7

Effect of process variables on film properties and defects

	Processing temp.	Screw and barrel wear	Resin selection	Die design	Screw design	Air ring and IBC
Gels	√	√	√	√	√	
Unmelts	√		√		√	
Gauge	√	√		√		√
Interfacial instability	√		√	√		
Haze lines	√	√	√	√	√	
Melt stability	√	√	√			√

Chapter 11

The cast film process

The flat die or cast film process is the alternative method for the production of thin polyolefin films. A typical machine is usually 2-3 m in die width with the web handling geared to run at a line speed of at least 400 mpm. Wider machines have been built capable of reaching line speeds of 600 mpm. Most machines will be equipped with multilayer coextrusion facilities. The largest applications are in the production of polypropylene and polyester films. Applications for polyethylenes are in stretch wrap films, embossed films for health care and backings for automotive and non-woven materials. The absence of orientation results in films better suited for thermoforming particularly with multilayer barrier films.

The process operates at high temperatures with polymers of relatively high melt flow rates. LLDPE resins of 3-5 dg/min melt index are used at temperatures ranging from 260-280 °C. These conditions result in low shear viscosity, which minimizes screw and barrel wear and reduces torque on the drive. **Figure 11.1** compares the shear viscosity of a cast film LLDPE (octene) at 280 °C to a blown film grade at 220 °C. The higher melt index and temperature should avoid the formation of sharkskin at high throughputs.

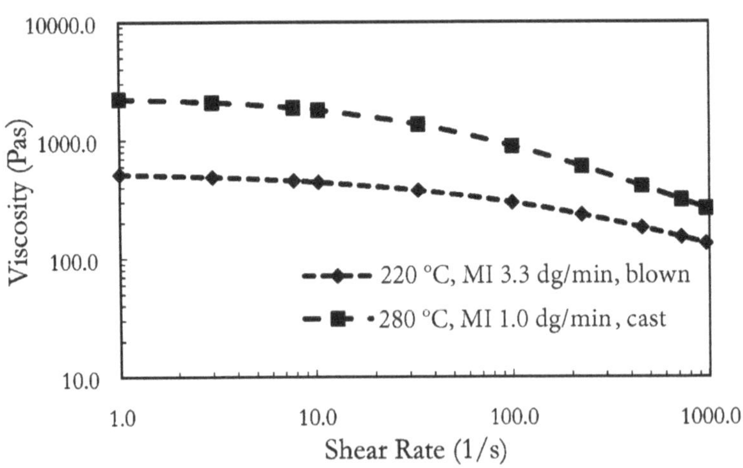

Figure 11.1

Comparative viscosity of cast and blown film grades of LLDPE

The heat transfer coefficient of the melt in air at a velocity of 10 m/s is around 40 W/m^2 °C whereas against steel it rises to 500-600 W/m^2 °C. This removes the cooling limitation inherent in blown film processes and allows cast film lines to operate at much higher outputs. Line speeds of 600 mpm are reported for LLDPE stretch film applications. High conductivity chrome plated steels should be used.

A typical extruder for cast film production is shown in **Figure 11.2**. The platform will be designed to accommodate satellite extruders for coextrusion. The extruder is placed on a moving carriage fixed on rails to allow movement to and from the chill roll assembly. A hydraulic lifting system to adjust the height of the die relative to the chill roll must be included. Sideways movement to adjust entry angle is also required.

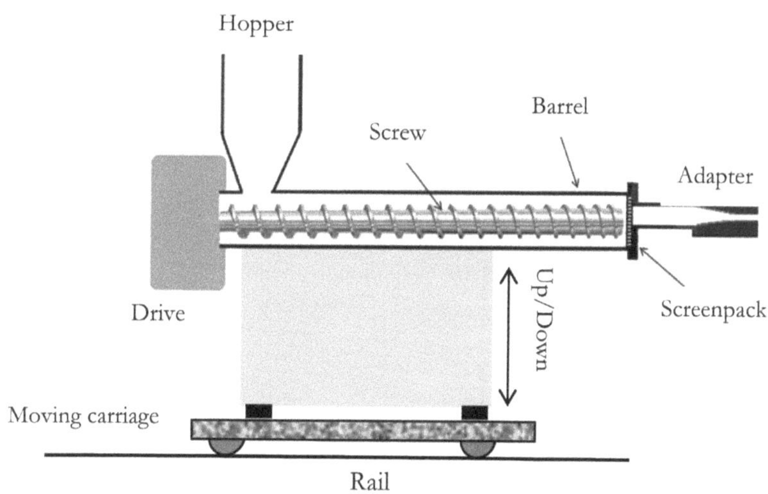

Figure 11.2

Extruder layout for cast film

Screw design must take into account the ability to achieve high output at high temperature. The development of high levels of viscous dissipation is, therefore, less of a problem. LD ratios of at least 28:1 are recommended to provide adequate time and pressure in the extruder. Compression ratios can be high with relatively shallow channels in the metering zone to maximize shear dissipation energy if required. Extruders with smooth bore feed sections are normally used. **Figure 11.3** shows the effect of the metering section channel depth on output for a 114 mm diameter screw at 150 rpm. Screw speeds are usually higher than with extruders designed for blown film.

Channel depth of the metering section will control the output from the extruder as shown in **Figure 11.3**. If the flight depth is too shallow output will drop and viscous dissipation may become excessive. The screw designer must, therefore, balance output with the optimum melt temperature for a specified range of polymers.

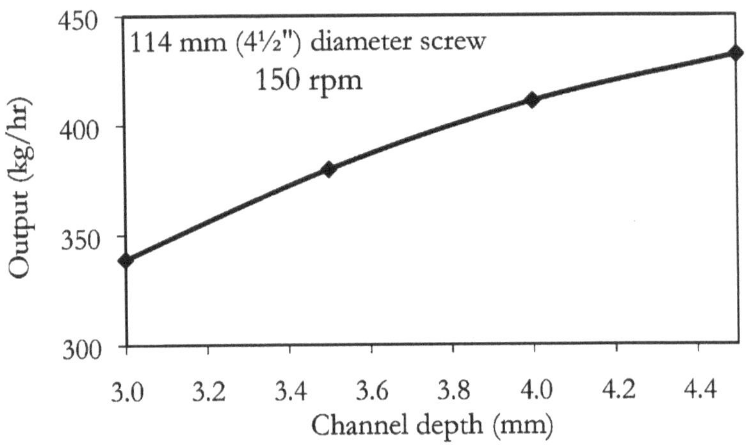

Figure 11.3

Effect of metering channel depth on output

11.1 Screws for cast film

A 114 mm diameter metering screw running with LLDPE was evaluated by computer simulation. The schematic of the screw is in **Figure 11.4**. Key design features are shown in **Table 11.1**.

Table 11.1

Screw design features

Screw component	Length	Depth (mm)
Length	30 D	
Feed zone	6 D	12
Compression zone	11 D	12 to 5.5
Metering zone entry	9 D	5.5
Maddock/fluted mixer	2 D	1.2 gap
Metering zone exit	2 D	5.5
Compression ratio	2.2	–

The back pressure was fixed at 10 MPa. The barrel temperature settings are shown in **Figure 11.4**. Two LLDPEs were run as shown in **Table 11.2**. At 100 rpm the outputs were 376 kg/hr. The exit temperatures for both resins were in the range of 279-283 °C. The higher melt index of the metallocene results in a slightly lower exit temperature.

Table 11. 2

Screw outputs

	LLDPE C_8	mLLDPE C_8
MI (dg/min.)	3.3	4.0
Density (g/cm^3)	0.917	0.917
Screw speed (rpm)	100	100
Output (kg/hr)	376	376
Temperature (°C)	283	279

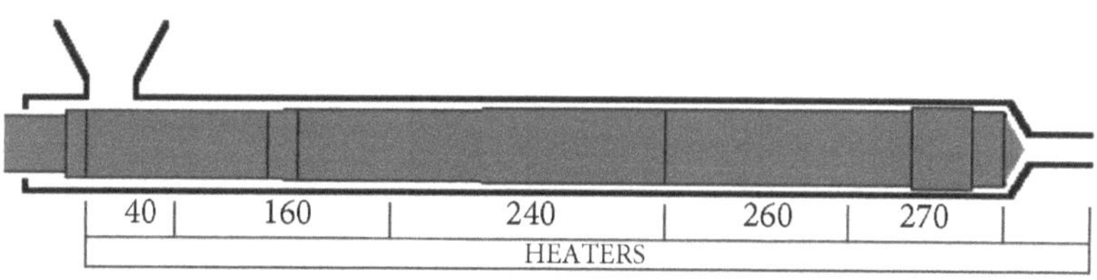

Figure 11.4

114 mm (4½'') Metering screw with mixer, 30:1 LD ratio

Figure 11.5 plots the temperature profiles for both polymers in the extruder. The extruder feed zone was set at 40 °C. Both resins melt at the same rate along the screw.

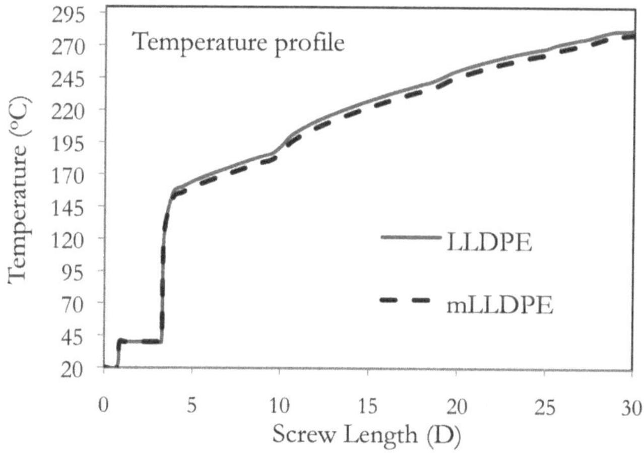

Figure 11.5

Temperature profiles along screw

The solid bed profile along the screw for the LLDPE is shown in **Figure 11.6**. The material is completely melted near the entry to the metering zone (17 D). This screw is able to extrude the two LLDPEs at the preferred melt temperature for this cast film process.

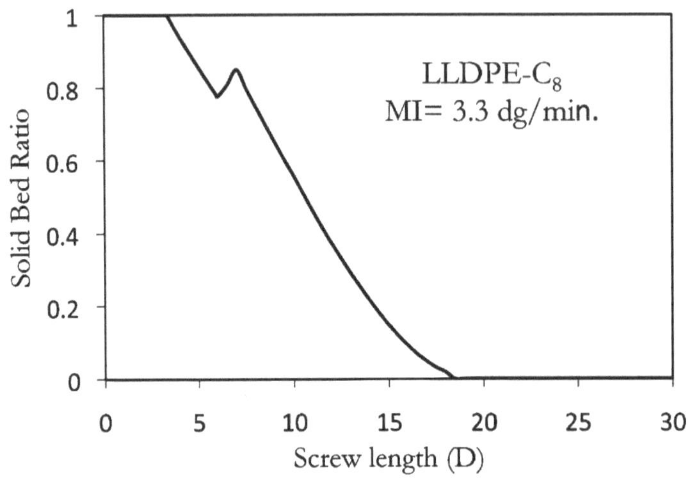

Figure 11.6

Melting profile in extruder

11.2 Flat dies

The underlying principle in die design is to avoid cross flows, which lead to uneven flow rates across the exit channel. The melt will take the shortest path from entry to exit and will flow preferentially down the middle. The key to good die design is to balance the pressure across its width to enable an even distribution of melt from the centre to the edges. **Figure 11.7** shows a typical coat hanger die with material flowing from the inlet port into the feed channel through to the restrictor. The coat hanger restrictor is designed to evenly distribute pressure across the width. The feed channel can be rounded or profiled in a tear drop, pear shape or other geometry. The final section is the die lip, which forms the melt into its final shape.

Cast film lines use wide dies and high line speeds. Good die design should be such that the polymer flow is streamlined with no chance of even tiny particles being trapped and degraded.

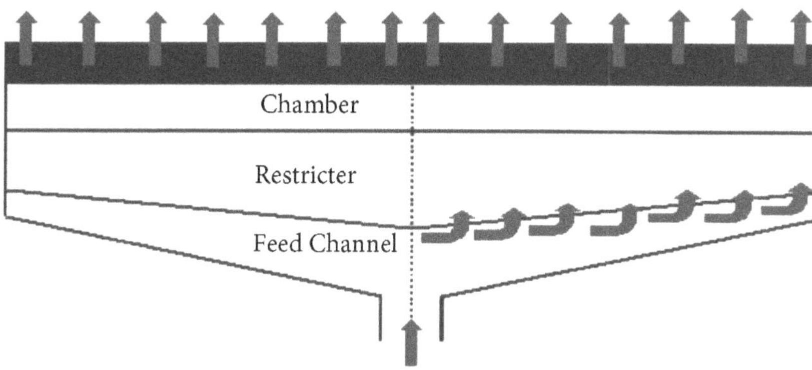

Figure 11.7

Balanced melt flow inside die

The functions of the die are as follows:

- To force the melt into a form approaching its final shape
- To maintain the melt at a constant temperature
- To distribute the melt at a constant pressure and at a uniform coating thickness
- To release the melt at the required width

Thickness is reduced as the melt is drawn and accelerates from the die onto the chill roll where it is rapidly solidified. **Figure 11.8** shows the schematic plan of a basic flat die design. The restrictor or distributor is designed in a coat hanger profile in order to encourage the melt to flow towards the edges of the die. Its

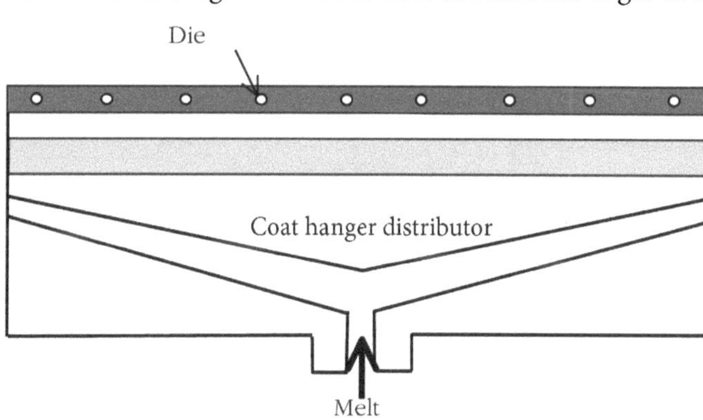

Figure 11.8

Plan view of coat hanger die

flow path will decrease from the centre to the edges to around one third to a half of its original surface area to form the coat hanger profile. The end plates bolted at the edges must be designed to ensure streamline flow and no hang-up (see **Figure 11.9**).

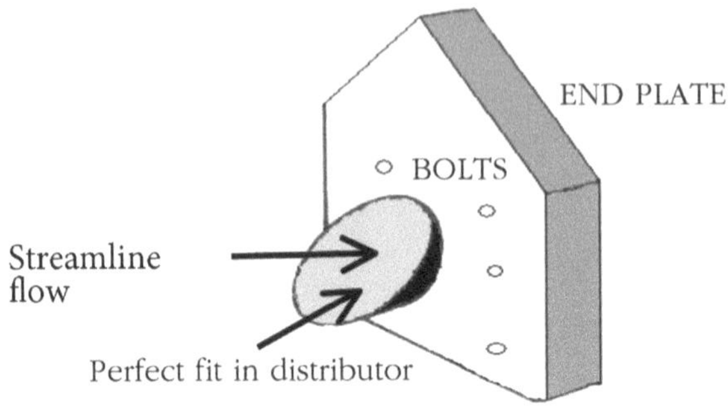

Figure 11.9

End plates

Figure 11.10 shows a cross section of the feed channel. The feed channel is normally designed in a pear or tear drop section and connects with the distributor. The shape of this channel can be critical in dies designed for feedblock coextrusion. Cheaper dies will use rounded channels, which are easier to machine from a single block of steel. These dies (T slot) are used for extrusion coating, but rarely for film and sheet applications.

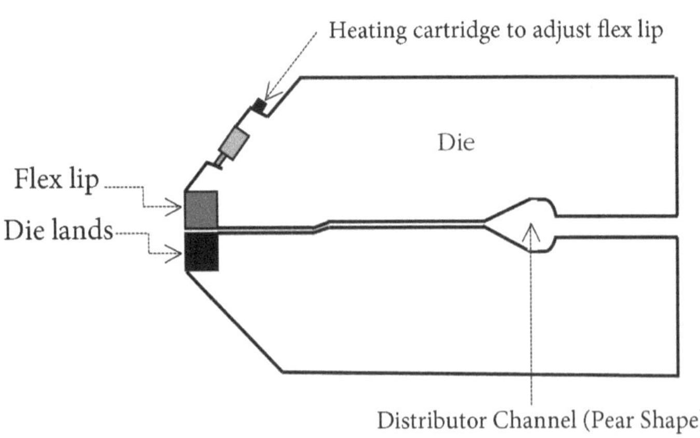

Figure 11.10

Cross section through a typical flat die design

Film thickness is adjusted by moving the die bolts which control the lip opening. Since the die controls film thickness uniformity it must be accurately machined of hard tool steel and incorporate highly polished surfaces. Its lower part has a V-shaped cross section so that it can be moved towards the chill roll. The lower portion of the die includes long hardened inserts or lands. The lands build up resistance to the melt flow and develop back pressure.

The die opening (gap) should be uniform across its width. Recommended die lip settings for PP and LLDPE are 0.5-1.5 mm.

To avoid corrosion problems with polymers such as EVA, ethylene acid copolymers (EAA) and ionomers it is recommended that all surfaces in contact with the melt are duplex chrome plated or nickel plated.

Computerised control systems have made it possible to automatically fine-tune the opening of the die lips to control thickness distribution.

These can provide the following benefits:

- Continuous real time measurements
- Data collection for process optimization
- Process control
- Reporting product compliance
- Raw material savings up to ~10%
- Increase in prime quality productions

With feedback from a scanning infrared thickness measuring gauge or beta gauge the computer can automatically correct for transverse machine thickness variations yielding films with a flat profile of ±2% variation. The designs incorporate flexible lips. The lips are adjusted by varying the heat input into die bolts, which determine the opening of the lips. The bolt temperatures are controlled independently with built in cartridge heaters and thermocouples. Simpler systems control temperature by varying the power (wattage) input into the heaters.

On heating the bolts will expand and the die lips will close. On cooling the bolts contract and the lips will open from the pressure of the polymer melt inside the die. If the melt viscosity is low, not enough internal pressure will be available to rapidly force the die lips to open. Delays of twenty minutes or more are possible before the required film thickness is delivered from the die. To accelerate the response push-pull systems are used. Spring back is faster with these designs some of which can adjust the thickness within three minutes.

Most automatic die designs incorporate permanent cooling with jets of air impinging on the bolts. Heating is practically continuous to compensate for the cooling. Full automatic thickness control is possible in both the machine and transverse directions. These controls are essential for efficient start-up and stabilization of coextrusion dies.

An automatic die with 2.1 m working width will be equipped with 70 expansion bolts across its width. As the expansion bolts are only subject to compressive loading, the flexible lips can be made to respond with precision to adjust the required thickness changes.

11.3 Die simulations

The die in **Figure 11.11** is used to produce a 20 μm film from typical LLDPE resins. The die is 3000 mm wide and the line speed is 360 mpm. Mass flow rate is 1400 kg/hr and all temperatures fixed at 280 °C. The LLDPEs described below are entered in this evaluation.

Type	MI (dg/min)	Density (g/cm^3)
LLDPE (C_8)	3.3	0.917
mLLDPE (C_8)	4.0	0.917

The flow path in the die is made up of three chambers from the coat hanger plus the adjustable lips. The flow chambers essentially taper down from the 3 mm profiled distributor to 1 mm. A 30 mm relaxation chamber, which opens out to 4 mm is added. This is reduced to 1.2 mm as the melt enters the lip section.

Feed channel - pear shape	L = 50 mm, W= 30 mm, Angle= 15°
Restrictor	110-60 mm from mid. to edge, gap= 3 mm
Die lips	Land = 5 mm, gap =1.2 mm

Figure 11.12 plots the percentage thickness distribution across the die for the 3.3 dg/min melt index LLDPE. The variation is ±1.5% from the mean. In **Figure 11.13** the shear stresses at the lips are in the range of 100-120 kPa and below the threshold for sharkskin formation for both resins. The output from the die is 4.7 kg/hr/cm of die width. This is more than double that achieved with a blown film process. The melt residence time ranges from 8-39 s from the centre to the edges.

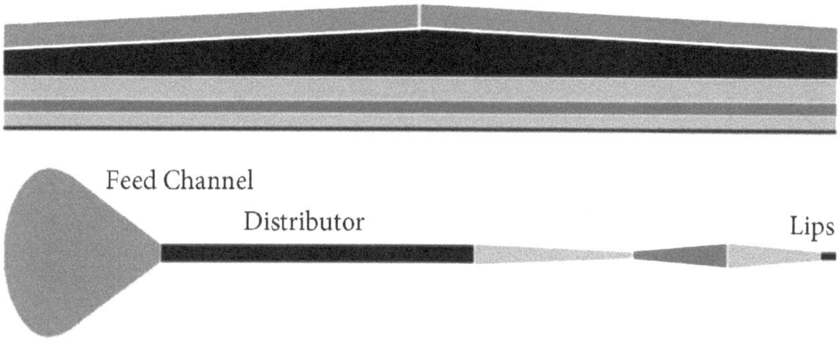

Figure 11.11

Coat hanger flat die

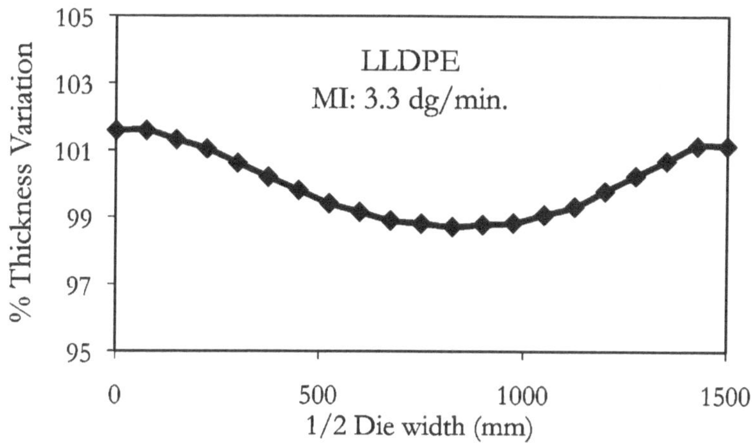

Figure 11.12

Flow distribution in coat hanger die

In **Table 11.3** a number of polymers are compared for shear stress at 280 °C in a flow channel as described below. This approximates to the die exit (lips) of a 1500 mm wide die. The mass flow rate is 700 kg/hr. (equivalent to 4.7 kg/hr/cm over the die width).

Die gap (mm)	1.2
Die width (mm)	1500
Die land (mm)	5.0

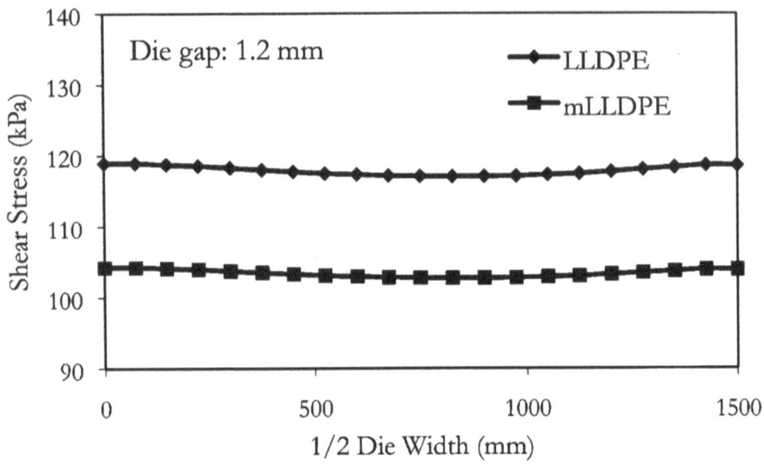

Figure 11.13

Shear stress across the die width
(LLDPE MI= 3.3 dg/min, mLLDPE= 4.0 dg/min)

The ionomer and the two octene LLDPEs can be safely run without sharkskin formation. The others should preferably be synthesized with a higher melt index (3.5-4.0 dg/min). The C_4 mLLDPE is produced by a slurry loop process and develops excessive shear stress of 216 kPa at these conditions.

Table 11.3
Shear stress in typical die lip section. Die gap: 1.2 mm

Material 4.7 kg/hr/cm	**Melt index (dg/min)**	**S/stress (kPa)**
Ionomer (Zn^{++})	4.0	70.2
mLLDPE C_8	4.0	106
LLDPE C_8	3.3	121
LLDPE C_4	2.8	178
mLLDPE C_6	3.0	197
mLLDPE C_4	3.4	216

It must be borne in mind that these computer simulations assume ideal flow environments. Operational conditions will be different particularly the effects of heat and pressure on the deformation of the die body and jaws. These may result in effects, such as "clam-shelling" where the die gap is wider at the centre of the die, which will cause flow rate variations across the width of the die. Large forces are generated in flat dies even at modest pressures. These forces will cause die plates and lips to deform and will require precise in-line die bolt adjustments to compensate. The increased residence time at the die edges may also cause a gradual build-up of degraded polymer in these regions, which may disturb flow. Other variables such as screw pulsation will also induce thickness variations adding to the complexity of the problem.

11.4 Chill roll cooling

Good design and maintenance are essential to hold a uniform temperature over the total surface of the chill roll. Double walled spiral rolls are utilized and designed to keep the temperature variation across the face to ±1-2 °C. A heavy inner shell supports a thin outer shell with helical flights for maximum heat transfer efficiency. High conductivity steel must always be used in the construction. The water flows through rotary unions connected to the two ends. Clean soft water must be used to keep the inner channels clean. Periodic flushing, cleaning and descaling are recommended to maintain the system at maximum heat transfer efficiency. The surface temperature should be higher than the environmental dew point of water to avoid condensation.

For high gloss films, the chill roll should have a peak to valley surface finish of 1-2 µm. The chill roll should be designed for rapid changeover to enable the use of different surface finishes if required. Rotary unions through which the cooling water flows should be designed for quick disconnects.

Figure 11.14 shows a schematic of a 500 mm diameter chill roll cooling layout. A secondary roll is added to further cool the web released from the primary roll. A 17 µm LLDPE film is produced at a line speed of 350 mpm with a die width of 1500 mm. The melt temperature is 280 °C and the chill roll temperature is set at 25 °C.

In **Figure 11.15** cooling along the chill roll surface is simulated. At a distance of 50 mm along the roll's surface the temperature drops to 50 °C in 8.6 milliseconds (ms). At 80 mm travel the film temperature is 25 °C; the same as the surface temperature.

Figure 11.16 compares the chill roll cooling effectiveness for the production of 20 µm films at three line speeds. The die width is 2000 mm. As the line speed increases the cooling curve shifts along the chill roll surface. All three plots reach the 25 °C surface temperature within 150 mm from contact with the chill roll. At 1000 mpm the mass flow rate is 1800 kg/hr. These calculations demonstrate the excellent cooling efficiency of the chill roll's steel surface and the ability to handle high extruder outputs.

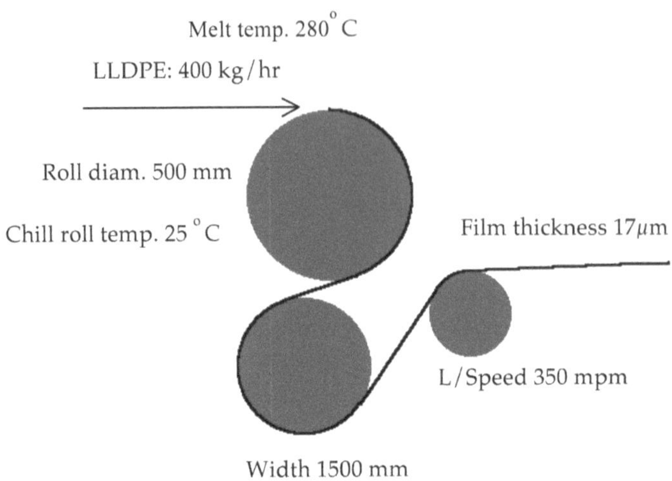

Figure 11.14

Chill roll layout

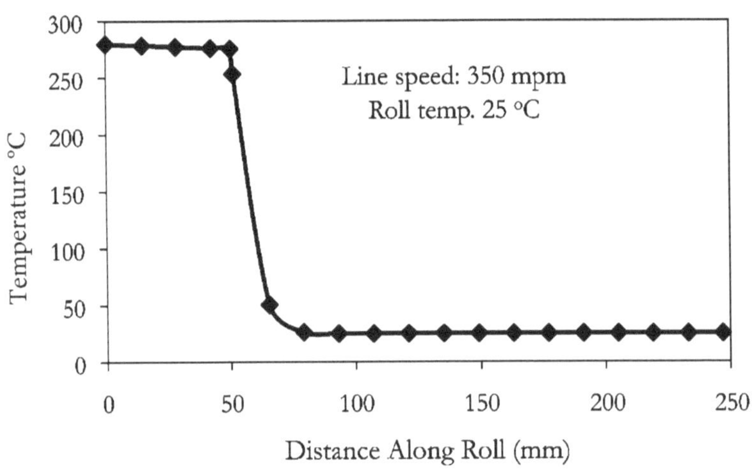

Figure 11.15

Film cooling along chill roll

Highly crystalline polymers such as polyamides require accurate control of the cooling process, with quite different conditions set according to the end use application. For maximum dimensional stability the melt should be slow cooled to enhance crystallization. For deep draw thermoforming a less crystalline (less perfectly formed morphology) film is preferred. It is important to remember that moisture will to some extent soften the film.

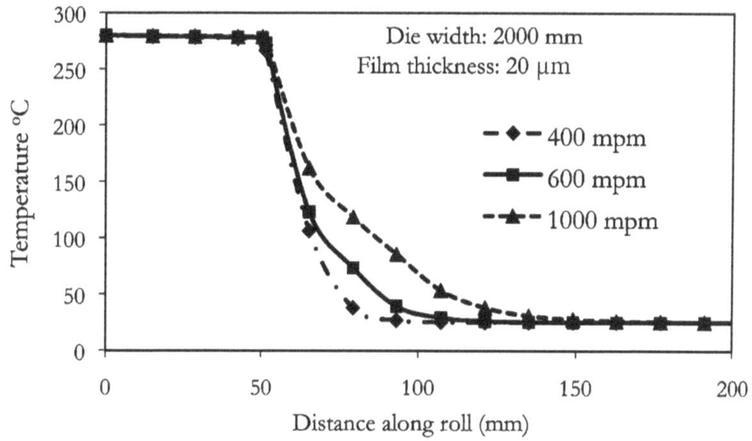

Figure 11.16

Film cooling at 3 line speeds

Suggested chill roll temperature settings for polyamide (PA6).

- Best thermoformability: 20-50 °C
- Dimensional stability: 75-85 °C

The next simulation was run for polyamide on a 1.5 m wide die with a throughput of 1800 kg/hr. Line speed was varied to produce 20 μm and 40 μm films. These were 720 mpm and 360 mpm respectively. The chill roll was set at 75 °C and the 400 mm diameter secondary roll set at 20 °C. The cooling regimen is shown in **Figure 11.17.**

The melts contact the primary roll at the extrusion temperature of 280 °C. At 258 mm from contact the 40 μm film has dropped to 102 °C and 77 °C for the thinner film. In spite of the doubling of line speed the 20 μm film cools faster. The resulting difference in crystallization rate may affect the properties of the final films. Both films, at 75 °C, feed onto the secondary roll set at 20 °C where they are released to the downstream haul-off equipment.

Figure 11.18 shows the effect of chill roll temperature on the optical properties of a 3-layer polypropylene coextrusion over a range of film thickness. The slower the cooling the higher or worse is the haze. The same trend is observed with transparency and gloss.

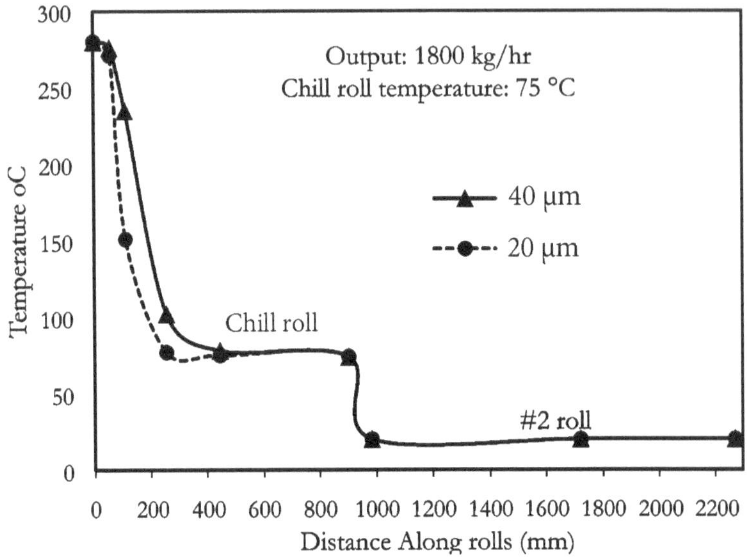

Figure 11.17

Effect of film thickness on cooling rate and crystallization

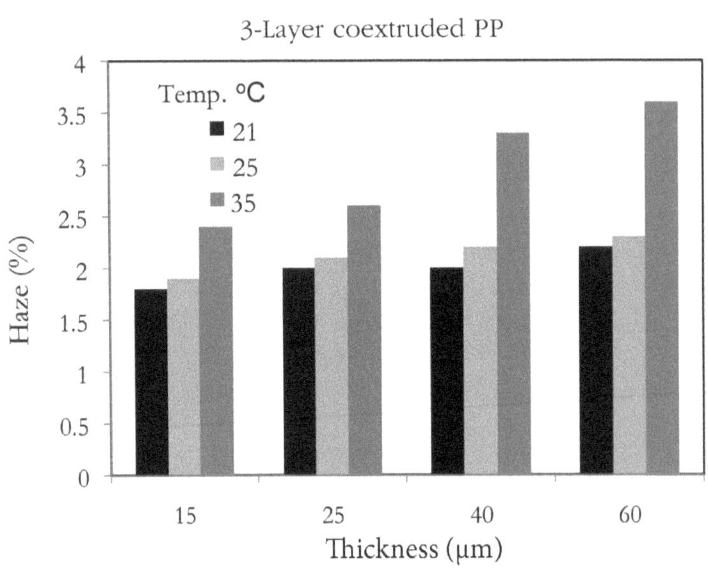

Figure 11.18

Effect of chill roll temperature on haze

11.5 Machine layout

Common dimensions for the cast film process are die widths of 1.5-3 m and thickness ranging from 10 μm to 250 μm. The corresponding thickness tolerances are approximately 1-5 μm. **Figure 11.19** shows a basic scheme for a cast film line. Two cooling rolls (primary and secondary) are included to maximize output. Some machines designed for biaxially oriented PP (BOPP) production are based on large extruder capacities designed to deliver outputs of around two tons/hr. Die widths of 4 m have been reported for 20 μm packaging films. Most of these lines will include coextrusion facilities.

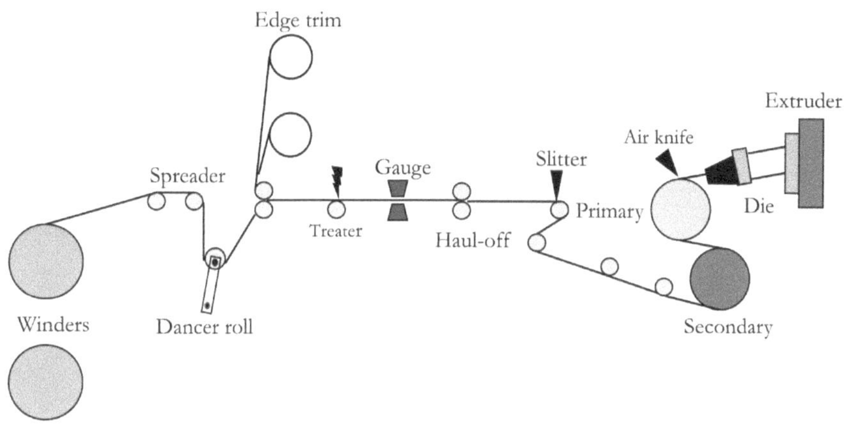

Figure 11.19

Schematic layout of cast film line

The melt stream from the die is formed into a thin flat film, which is cast and drawn on the cooled metal roll. Double shell spiral contacting baffle designs can enable the reduction of the outer shell thickness from 19 mm to 9.5 mm. This improvement in thermal transfer can increase output by some 10%. The chill roll temperature is controlled according to the requirements of the polymer and application. The degree of crystallinity developed by the polymer is very much dependent on its rate of cooling. The faster the cooling rate the lower the crystallinity. Both rolls are usually of the same diameter, however some machines may have a smaller secondary cooling roll. On the second roll the film is cooled down to room temperature. Additional annealing rolls can also be added to minimize curl and post shrinkage.

To avoid air inclusions between the melt and the first chill roll surface, the melt is pressed to the roll by an air knife operated with high velocity air from a compressor. The air must be filtered to prevent dust from being blown onto the melt.

To minimize neck-in at the edges, devices that pin the edges of the film against the chill roll are used. Air-knives have to be precision engineered and optimally positioned in order to get the best results. Poor gauge control and transparency can result from poorly designed and positioned air-knives. The set-up for positioning the air knife and pinning the edges to the chill roll is shown schematically in **Figure 11.20**.

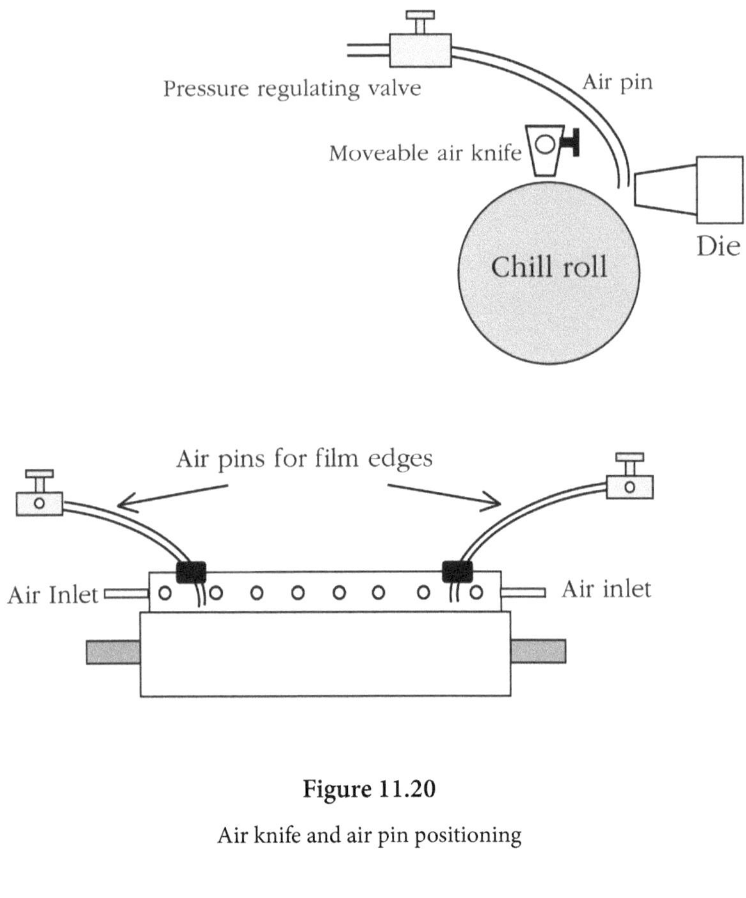

Figure 11.20

Air knife and air pin positioning

Some processors prefer the vacuum box system. This technique draws the melt to the chill roll instead of pressurizing it against the metal surface. These devices are claimed to practically eliminate air entrapment between web and roll and provide extra cooling. Adjustment of the vacuum is important as this can affect melt relaxation time and consequently film properties. The pressure should be around 2-4 cm of water. Vacuum boxes are costly, but are claimed to work very well at low gauges. The vacuum can also eliminate fumes from the melt and may help to reduce plate-out on the chill roll. Machinery vendors claim that vacuum boxes are mandatory for line speeds of 300 mpm and above.

The edge beading is trimmed away and collected on winders (rolls) further downstream. It is important that the trim be kept clean for recycling. The wider the film the less significant the edge trim becomes as a percentage of total production. The chopped trim can be conveniently recycled into the middle layer of an ABA or ABC coextrusion set-up.

The next station in the line is the thickness gauge followed by the corona treater. The treatment energy for LLDPE film should be at least 20-28 W/cm^2, depending on slip concentration and electrode design. Cast film lines will usually operate at haul-off speeds that can be five times higher than blown film lines.

In cast film production control of thickness is very critical. The wound roll must be perfectly flat and free of any gauge bands. To guarantee film flatness, three precautions should be taken:

- The die should be equipped with an automatic thickness control system.
- The surface temperature of the chill rolls should be kept within a range of ±1.5 °C.
- Between the secondary chill roll and the edge trim slitter an oscillator roll should be installed, which traverses the film in the TD at an adjustable traverse of 10 to 100 mm.

Recent developments in computer controlled web handling units makes scrapless winding of film possible at speeds of 600 mpm. Thickness can range from 10-150 μm and roll diameters up to 1000 mm can be produced without wrinkles, fold back or film elongation during roll changes. These systems reduce the transfer time on centre winders from 12-60 s down to 2 s. These design improvements makes it possible to achieve very high levels of productivity in applications such as stretch film and embossed diaper (nappy) films. Coupled with feedblock coextrusion facilities the cast film process has proven to be very flexible in the range of polymers that it can process, e.g. PET, APET (amorphous PET), PA. PP, EVOH etc. and its use is increasing in many speciality film applications.

11.6 Drawdown and neck-in

In flat die extrusion the molten polymer is forced through a very wide, but thin gap of approximately 1-1.5 mm. The melt is drawn and rapidly quenched on a chill roll or in some cases immersed in a water bath where it rapidly solidifies. Due to its viscoelastic properties the melt will swell as elastic shear forces are relieved and the constraints of the die are relaxed. The flat melt stream will develop an edge that is thicker than the rest of the film. The build-up at the edge is a consequence of the neck-in of the melt. This will depend on the elongational viscosity of the polymer and the distance between the die lips and the chill roll. These thicker edges must be trimmed off as discussed earlier. A sketch of the process is in **Figure 11.21**.

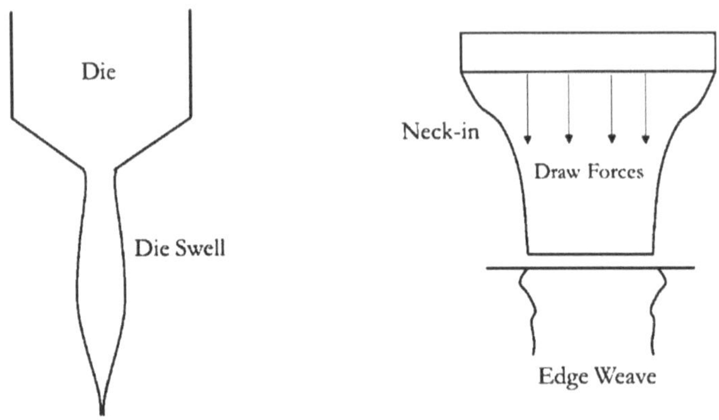

Figure 11.21

Neck-in and edge tear at die exit

If the polymer has a high elongational viscosity (LDPE, EVA etc.) high stresses result in the melt causing edge tear and ultimately total melt break. With tension thinning polymers (LLDPE and PP) the elongational viscosity is low and high draw rates are achieved (i.e. rates of extension), but edge beading will be high. This is counteracted by the use of very short air gaps and high stretching rates, which minimize edge trim.

A frequently encountered form of instability in very high draw rate processes is the phenomenon known as draw resonance. Draw resonance is the periodic fluctuation in the air gap at constant stable extrusion and web speed. The fluctuations are non-uniformity of web thickness in the machine direction. Draw resonance is rationalised in the following way. As the melt is drawn, its viscosity is reduced as the applied strain increases. The extrudate dimensions are continually changing in the air gap with melt velocity. Just before the freeze line the melt is at its thinnest and consequently at maximum strain. If viscosity now drops just before the melt freezes and becomes dimensionally stable, the extrudate just upstream is at a lower strain and will consequently have a higher viscosity. This viscous and thicker material will then flow to the freeze line leaving thinner extrudate behind it. A cyclic variation is imposed and the extrudate will now continuously resonate from thick to thin sections. Short air gaps as used in cast film production will usually avoid draw resonance occurring. In the extrusion coating process where air gaps of 150-200 mm are used this phenomena becomes problematic with linear polyethylenes and polypropylene.

11.7 Differences between blown and cast films

In comparison to blown film, cast film processing conditions differ in the following respects:

- The extrusion temperature is 40-60 °C higher
- The melt cooling rate is around 10 × faster (rapid quench vs. air cooling)
- Flexible lips are easily adjusted to minimize thickness variation
- No turbulence from cooling air
- Additional cooling possible with air-knife or vacuum box
- There is no cross direction (TD) orientation
- Deckling to control width can cause polymer degradation behind deckles

These conditions lead to the following properties differences compared to blown film:

- Lower stiffness
- Better gauge control
- Higher softness and handle
- Excellent transparency and gloss (reduced crystallinity)
- Lower barrier (lower crystallinity)
- Balanced strength and tear resistance in MD and TD
- Low shrinkage

In coextrusion the cast film process based on a combining feedblock has the following advantages:

- Flexibility in layer ratios
- Ultra-thin layers
- Better overall gauge control (via flex lip)
- Greater facility for varying layer ratios
- Easier cleaning and maintenance

- Lower cost of feedblock
- Layer multiplication (microlayers)

The disadvantage however is that the layers have to flow from the feedblock through the die and into the lips. This increases the probability of layer rearrangement due to viscoelastic and geometric effects. In the annular dies used in blown film there is no layer interface in contact with the die walls and rearrangement will not occur.

Chapter 12

Coextrusion

Coextrusion is the simultaneous extrusion of two or more polymers to form one multilayer composite. The process produces thermoplastic films and sheet, which benefit from the synergy provided by all the component parts of the structure. For films and laminations used in packaging the characteristics of critical importance to the user are:

- Heat-sealability and hot tack
- Stiffness or flexibility
- Surface properties
- Impact strength
- Tear resistance
- Dimensional stability
- Balanced properties in MD and TD
- Water vapour transmission rate (MVTR)
- Gas (O_2 /CO_2) transmission rate (OTR)
- Aroma transmission
- Scalping (absorption of flavours, nutrients, etc.)
- Transparency (optional)

Additional attributes that are also important to the processor must also be considered. These are:

- Cost
- Use of recyclates and in-house scrap
- Interlayer adhesion
- Downgauging (resource reduction)
- Low extensibility
- Process stability (melt strength, thermal etc.)
- Extruder output

- Screw/barrel wear

Coextrusion is the technique used to combine into one structure and one machine pass as many of these attributes as possible.

12.1 Typical coextrusion problems

Practical experience over many years has shown that the melt flow behaviour of coextrusions can be unpredictable and much more complex than originally assumed. A number of problems can occur when two or more non-Newtonian melt streams flow in a multilayer flow field. The most common defects are illustrated in **Figure 12.1**.

Poor layer uniformity is common in feedblock systems and is caused by the tendency of one layer to encapsulate another. This phenomenon is the most difficult to eliminate.

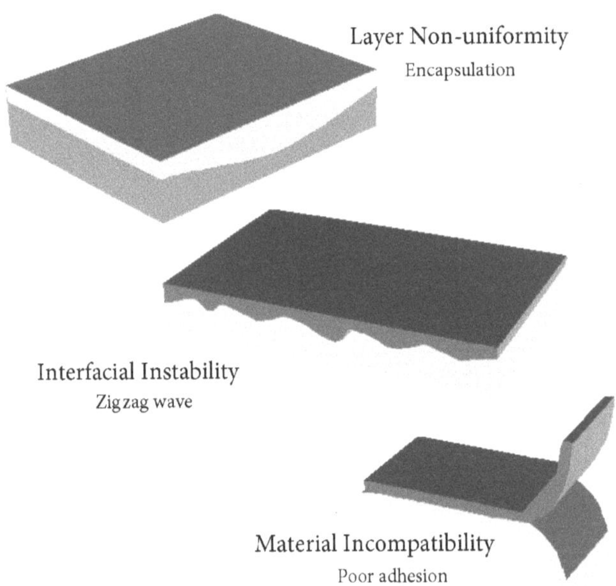

Figure 12.1

Common coextrusion defects

Interfacial instability (zigzag wave) occurs when excessive shear stress develops at interfaces. Widening channels, raising temperature and reducing mass flow rate will usually minimize or eliminate this defect. More intractable forms of interfacial disturbances can also occur, which will require other adjustments, such as balancing of flow velocities at interfaces.

Material incompatibility can be overcome by incorporating compatible polymers or introducing tie-layers and/or compatibilizers. The problem can occur with all types of dies and processes.

12.2 Melt encapsulation

There are three basic die shapes in extrusion processes:

- Slit or slot dies
- Cylindrical dies
- Annular dies

In slit and cylindrical dies the layer interface will be in contact with the channel wall, which may lead to one layer leaking towards the walls and displacing the adjacent layer. This will lead to layer rearrangement and encapsulation. In annular dies the interface is never in contact with the walls and encapsulation caused by leakage of a layer towards the walls will not occur, i.e. blown film dies do not suffer from this type of layer rearrangement.

The tendency for a lower viscosity melt to encapsulate a higher viscosity fluid is demonstrated in **Figure 12.2**. The phenomenon is usually referred to as viscous encapsulation. The longer the flow path the greater the degree of encapsulation

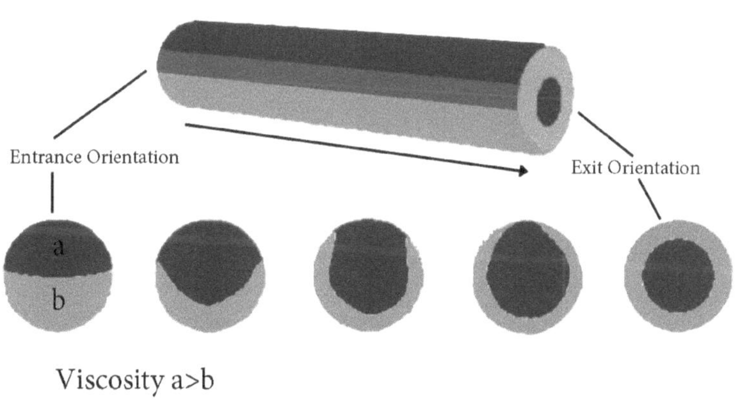

Figure 12.2

Encapsulation in coextrusion

as illustrated by the five steps in **Figure 12.2**. The lower viscosity layer (b) gradually envelops layer (a) as the melt progresses through the channel. A coaxial distribution can be observed when the LD ratio is >100.

This predicts the outcome but does not explain when or where it will occur. The obvious solution is to keep the flow paths as short as possible and avoid viscosity mismatches. Using multimanifold dies where the melt streams are brought together just before the die exit is one solution. However, wide flat die multimanifold systems are very costly and difficult to set up and consequently the cheaper and more flexible combining feedblock designs are usually preferred. Tubular dies avoid this problem as mentioned above but are very cumbersome when more than two layers are required.

For many years the misconception persisted that interlayer thickness variation, i.e. the beginning of encapsulation, was caused solely by viscosity differences. However, work by Dooley[1] (Dow Chemicals) and others has shown that the encapsulation phenomena can occur with a single polymer in the coextrusion die. This research showed that layer rearrangement could also be caused by secondary flows originating with viscoelasticity and geometry. These experiments demonstrated that highly viscoelastic polymers such as LDPE and polystyrene were more prone to encapsulation than for instance the less viscoelastic polycarbonate. The problem is aggravated in long dies where elastic forces continue to affect structures throughout the flow domain. The phenomenon has been linked to the following effects:

- Melt elasticity
- Normal stresses
- Memory

Other researchers have claimed that the phenomenon is principally linked to flow geometry where for instance in a particular flow path one layer may find an easier passage. Once this happens localized thickening will take place leading to depletion elsewhere causing layer rearrangement.

Since there is no mathematical model or theory to predict exactly when and how encapsulation will occur, die designers have over the years developed partial solutions to solve the problem. It is now customary to design modular feedblocks with replacement inserts to divert the flow as required. Melt profiling is virtually by trial and error and the problem becomes increasingly complex as the number of layers increases.

[1] Dooley, J., Hilton, B.T., Layer Rearrangement in Coextrusion, Plastics Engineering, **1994**, 50, 2, 25

Figure 12.3 shows a typical feedblock profiling technique used to compensate for the encapsulation of the middle layer in this example. By profiling the middle layer, in the feedblock, into a concave geometry ("bow tie") the encapsulation tendency by the outer layers is minimized. The middle layer is encouraged to flow to the edges of the die.

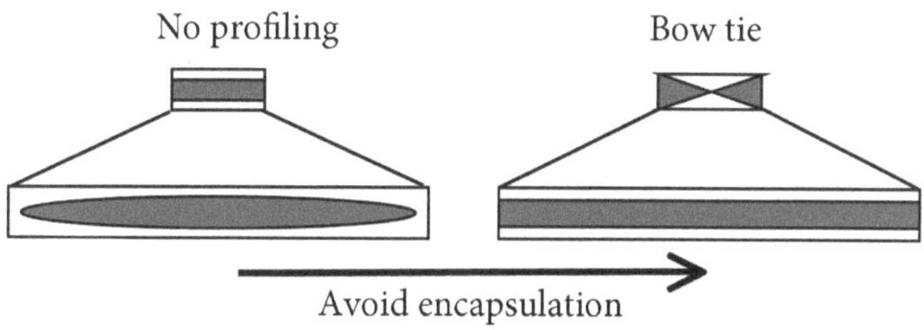

Figure 12.3

Feedblock profiling

It must be remembered that the problem is not simply viscosity related as discussed earlier, but also due to other poorly understood phenomena. Therefore, it is possible for the profiled melt to rearrange itself to an unpredictable shape as it leaves the feedblock and flows into the complex geometry of the die where squeezing effects and other rapid velocity and elongational changes take place through the flow channels.

12.3 Interfacial instabilities

In a simple coextrusion flow field as shown in **Figure 12.4** excessive shear stress at the interface can cause instabilities. The instability when it occurs will manifest itself by an increase in haze, as measured by narrow angle light scattering and the formation of zigzag waves. Reducing pressure in the die will usually eliminate these instabilities.

Zigzag wave formations on a typical tubular blown film bubble are illustrated in **Figure 12.5**. Two types of wave can be formed as shown. The higher amplitude wave with a lower frequency can result in intermittent layer flow. Its origin is elusive, but could be initiated in the region where the polymers merge. Excessive melt acceleration at interfaces and the layer ratio can cause low frequency wave instability.

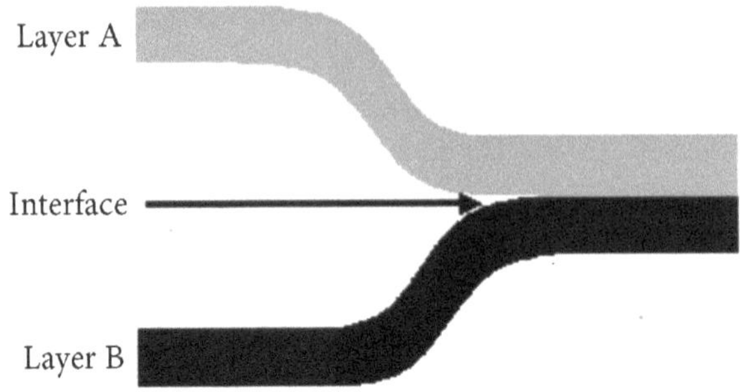

Figure 12.4

2-Layer coextrusion flow field

Two models are used to explain wave formation:

- At a critical shear stress, interface adhesion drops causing slippage
- Instabilities are caused by the development of normal forces at the interface

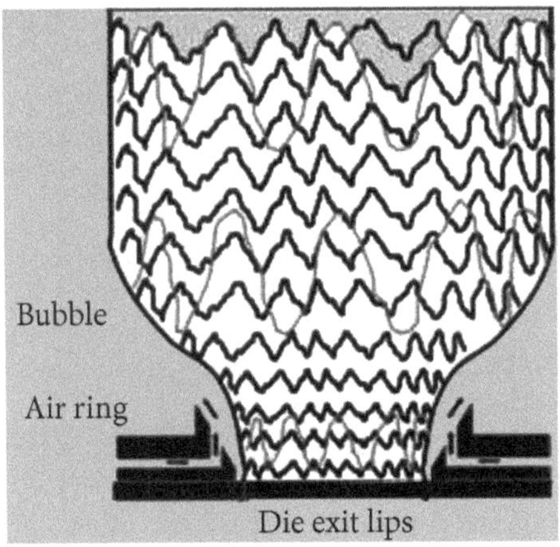

Figure 12.5

Zigzag wave formation

Irrespective of the model used observations have shown that the onset of interfacial instabilities can occur at a shear stress of approximately 30 kPa. If the shear stress reaches 50-60 kPa interfacial instability is highly probable. These critical values are subject to debate by theoreticians and should, therefore, be applied with some caution.

In a simple monolayer flow field shear stress will increase from the middle to the walls of the channel. An example is shown in **Figure 12.6** where the shear stress across a pipe of 4 mm diameter with a length of 25 mm is plotted. The LDPE is flowing at a rate of 500 kg/hr at 300 °C. Shear stress is reduced from the wall to the centre of the flow domain. At the wall the shear stress reaches 140 kPa.

This observation must be considered in coextrusion where at least one interface will be present between layers. The closer the interface is to the die walls the higher will be the shear stress leading to possible instability. This can be problematic with outer skin layers where the interface is very close to the die walls. To minimize this effect a low viscosity melt should be placed as the outer layer close to the walls of the flow channel.

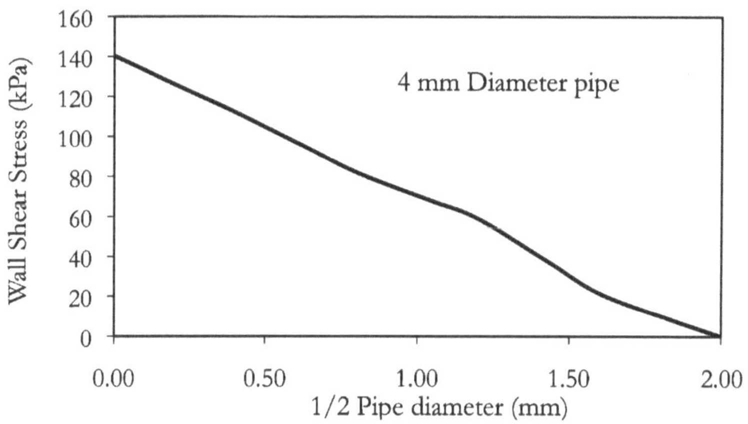

Figure 12.6

Shear stress across a channel from wall to middle

A symmetrical tubular film coextrusion based on LDPE and HDPE was simulated. The final die section is configured in **Figure 12.7**. The die diameter is 250 mm. The three layers are combined 10 mm from the exit and flow through a die gap of 1.5 mm. Melt and wall temperatures were fixed at 240 °C. The total mass flow rate was 185 kg/hr.

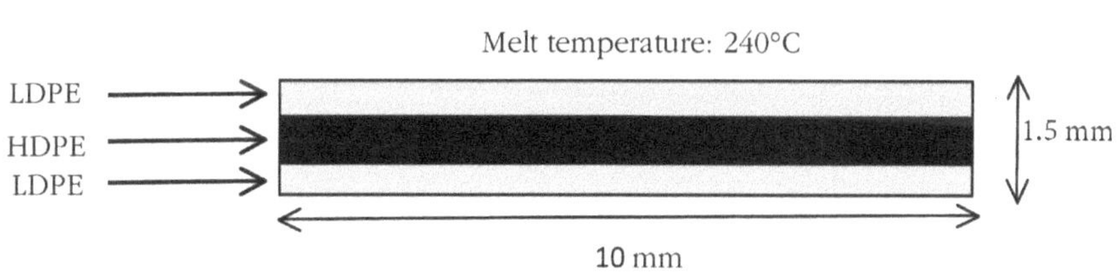

Figure 12.7

3-Layer coextrusion with narrow gap

The basic properties and entered flow rates of the two polymers are shown in **Table 12.1**. The outer layers (LDPE) account for 10% of the thickness per side. The shear stress data at the walls and interface calculated by the software are summarized in **Table 12.2**. From these calculations the shear stresses are below the thresholds that are likely to lead to instabilities or melt fracture related defects.

Table 12.1

PE Resin entry

Polymer input	Layer #1	Layer #2	Layer #3
	LDPE	HDPE	LDPE
Input (kg/hr)	18.3 (10%)	146.4 (80%)	18.3 (10%)
Melt index (dg/min.)	2.0	0.45	2.0
Density (g/cm^3)	0.923	0.945	0.923

Table 12.2

Interfacial shear stress. 1.5 mm die gap

Result (LDPE/HDPE/LDPE)	Value (in/out)
Shear stress at walls (kPa)	39.1/39.1
Shear stress at interfaces (kPa)	26.6/26.6

The LDPE is now replaced by a POE type metallocene resin (mPE) in order to take advantage of its superior heat-seal properties. This resin has a melt index of 1.1 dg/min and a density of 0.902 g/cm^3 and is characterized by very narrow molecular weight distribution. Mass flow rates and temperatures were unchanged. The results in **Table 12.3** show that much higher levels of shear stress have now developed inside the die.

Table 12.3

Shear stress in die. 1.5 mm die gap

Result (mLLDPE/HDPE/mLLDPE)	Value (in/out)
Shear stress at walls (kPa)	151/151
Shear stress at interfaces (kPa)	100/100

Shear stress of 151/151 (in/out) kPa at the walls will possibly result in sharkskin and 100 kPa at the interfaces will lead to severe interfacial instability. Redesign of the die is now required in order to run this coextrudate. The simplest option was to widen the die gap to 2.5 mm. The resultant shear stress data is summarized in **Table 12.4**.

The wall shear stress is reduced to 82.8 kPa and is well below the threshold for the onset of sharkskin. However, the interface shear stress is still too high at 54.9/54.9 kPa.

Table 12.4

Shear stress in die. 2.5 mm die gap

Result (mLLDPE/HDPE/mLLDPE)	Value (in/out)
Shear stress at walls (kPa)	82.8/82.8
Shear stress at interfaces (kPa)	54.9/54.9

The next step was to increase the ratio of mLLDPE to HDPE in the system. This will push the interface away from the walls to a region of lower shear stress. Total mass flow rate was kept at 185 kg/hr and the outer layers were increased to 15% per side as in **Table 12.5**.

Table 12.5

PE Resin entry

Polymer input	Layer #1	Layer #2	Layer #3	Total
	mLLDPE	HDPE	mLLDPE	kg/hr
Flow rate (kg/hr)	27.75 (15%)	129.5 (70%)	27.75 (15%)	185

Table 12.6 shows that moving the interface away from the walls by increasing the thickness of the mLLDPE outer layers reduces the shear stresses at the interfaces. The 44 kPa interface shear stress is now borderline.

Table 12.6

Interfacial shear stress. 2.5 mm die gap

Result (mLLDPE/HDPE/mLLDPE)	Value (in/out)
Shear stress at walls (kPa)	79.0/79.0
Shear stress at interfaces (kPa)	44.0/44.0

The simulation clearly shows the difficulties that may be encountered with the use of very narrow molecular weight distribution (MWD) polymers. Widening the die gap further is not recommended since at 2.5 mm the gap is already very wide and will lead to excessive drawdown in the machine direction (MD) and will also result in poorer control of thickness distribution from the die. Increasing melt temperature will cause difficulties in bubble cooling. Finally, reducing the throughput of the die will reduce pressure, but is the least desirable option leading to productivity loss. The most practical solution is to blend the mLLDPE with LDPE or EVA, which will significantly reduce the overall pressure in the die.

Another option, which adds cost, is to blend with fluoroelastomer processing aids. These will generate slip at the die walls thus reducing pressure.

12.4 Blown film coextrusion

The development of zigzag waves and interfacial instabilities is more prevalent with tubular dies due to the higher pressures developed with low melt flow polymers and the lower temperatures required for melt stability and facilitating bubble cooling. Low temperatures and high polymer shear viscosities will increase viscoelasticity (memory) and cause the melt to be more susceptible to entry angles and excessive stress at interfaces. In contrast flat die processes run at higher temperatures with "easier" flow resins have virtually no cooling limitations and are consequently less prone to these interactions.

In a tubular die the melt streams are preferably led from a port block into the die, which feeds the individual melt streams from the extruders to their corresponding flow channels. These are a series of concentric spiral cut mandrels, which distribute the melt streams independently. These overlapping layers are combined around a central mandrel just before the exit. One advantage is that the layer rearrangement phenomenon typical of feedblock designs does not occur with annular dies. **Figure 12.8** shows a typical 5-layer design. This shows each layer as a virtually independent die within the whole. Thickness is adjusted independently for each layer.

Coextrusion tubular dies can be very cumbersome and are difficult to calibrate, purge and maintain. Unlike the simple combining feedblock, these are dies within dies and become excessively large and heavy as the diameter increases from the inner mandrel with each addition to the outermost layer. This increase in diameter exposes the melt to a greater surface area, which increases pressure (shear stress) and creates the potential for interfacial instabilities and degradation. The problem is aggravated as the number of layers increases. However, spiral mandrel dies with up to 7-layer coextrusions are produced and used successfully.

A very high barrier film can be produced in a 7-layer set-up by exploiting the natural affinity between PA and EVOH. The core layers would be PA/EVOH/PA. This combination is virtually inseparable. The remaining 4 channels will accommodate the 2 tie-layers and polyolefin skins. The EVOH will reduce OTR and the polyamide will improve thermoformability, flex crack and toughness. The absence of a metallic layer permits microwaving. These films can be competitive with aluminium foil laminates and metallized films in packaging perishable foods.

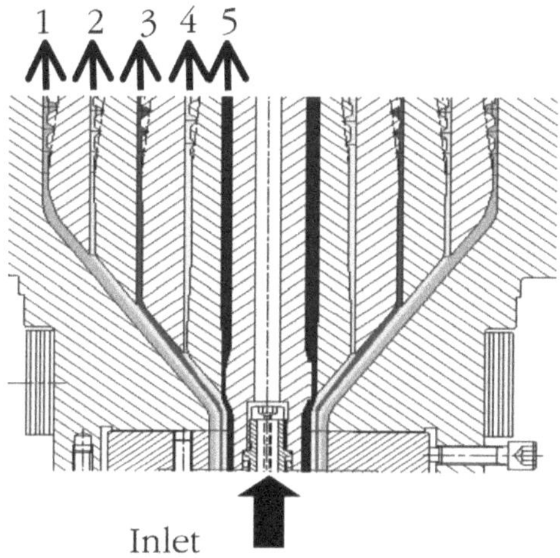

Figure 12.8

5-Layer tubular coextrusion die

Figure 12.9 shows a number of possibilities of blown film coextrusions with different arrangements of extruders and dies. Blow-up ratios (BUR) are in the range of 1.5-2.5.

To avoid gauge bands on the finished rolls oscillating dies are used. However, as dicussed in a previous chapter, these have a number of disadvantages:

- Increased proneness to leaks at the seals of the rotating mechanism
- Stripping and reassembly for maintenance is time consuming
- Purge times can be excessive
- Die height typically increased by 1 m

Therefore, fixed dies with rotating haul-offs are preferred. The nip rolls will have 360° rotation and are equipped with turning bar systems that guide the layflat down to the winder placed adjacent to the extruders.

The pancake die was developed to overcome the limitations of the spiral mandrel die in the blown film coextrusion process. This design is based on cutting the spiral channels on the surface of a flat plate. The melt distribution occurs in a plane that is perpendicular to the axis rather than in the usual concentric flow direction. The melt streams are fed to each layer from the side of the plates and split several times as they flow along the channels towards the centre of the die. **Figure 12.10** shows an example of a pancake section cut with multiple spiral grooves.

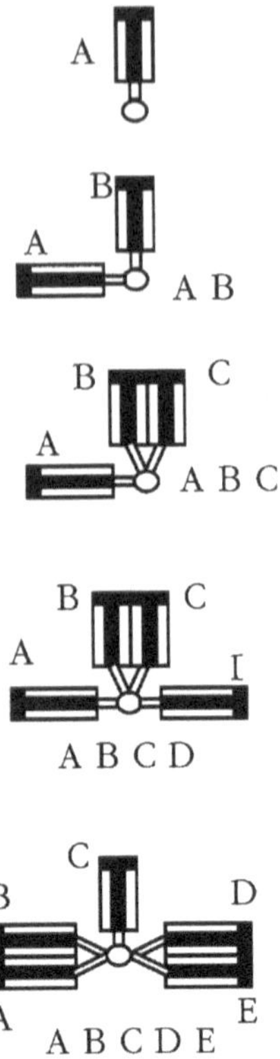

	STRUCTURE	MATERIALS
	A	ALL PEs
	AB or ABA	A=PE etc. B=PE etc.
	ABC or ABCBA	A=PE etc. B=PE or tie-resin C= PE, PA, EVOH
	ABCBD or ABCBCBD	A=PE, PA B= tie-resin C=PE, PA, EVOH D=PE, PA
	ABCDE or ABCBDBE or ABCDCBE	A & E= PE, PA B & D= tie-resin C= PE, PA, EVOH A & E= PE, PA B= tie-resin C= PE, PA, EVOH D= PE, PA, EVOH

Figure 12.9

Tubulaar film coextrusion layouts

Figure 12.10

Pancake section with multiple channels

The advantage of this design is the ability to stack and bolt these plates on top of each other to create the multilayered structure. Each layer is added sequentially to the previous layer and the combined melt streams flows up the die towards the exit. This design is versatile as it is relatively easy to increase or decrease the number of plates stacked together. Another advantage of the pancake design is the ability to heat insulate each segment in order to maintain temperature differentials between layers. This is more difficult to achieve with the conventional vertical spiral mandrels. Coextrusions with up to 11 layers can be produced using these pancake designs. Increasing the number of plates further will usually develop excessive pressures leading to interfacial instabilities, leakage at joints and overloading of the extruder drives. In addition controlling layer thickness distribution becomes more challenging.

However, recently it has been shown that by designing very thin plates it is possible to increase the number of layers to around 30. Developments are continuing to produce nanolayers in blown films based on using feedblocks and layer multipliers in combination with proprietary dies able to overcome the high pressure and excessive residence time limitations.

An interesting option is the idea of blocked films. This technique allows the layflat to block and form one single sheet. Property improvements are claimed by this process for any given combination. **Figure 12.11** shows an example of 7-layer layflat blocked to produce a 14-layer barrier film. The PA/EVOH layer is doubled and forms the core barrier layer. The usual tie-layers are used to combine the barrier structure to the PE. The addition is a tacky layer of EVA as the inside skin. A simpler solution is to apply the EVA as the inner layer. This is typically

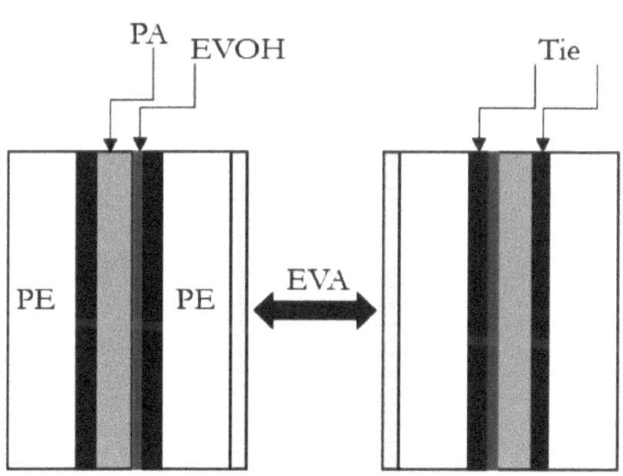

Figure 12.11

Example of 14-layer blocked barrier film

10 μm thickness. It is important to remember that the layer thicknesses must be adjusted for the doubling up process. As an example if the application specifies 10 μm EVOH, this layer is now reduced to 5 μm in the coextrusion.

12.5 Coextrusion systems for flat dies

There are two basic coextrusion systems in commercial use for flat dies. The most common is the combining feedblock system see **Figure 12.12(a)**. The second concept is based on the construction of two or more manifolds in the one die. The basic layout is shown in **Figure 12.12(b)**. The melts can be combined outside or just inside the die. The third option is to add a feedblock to one or both of the manifolds thus increasing the number of layers, **Figure 12.12(c)**.

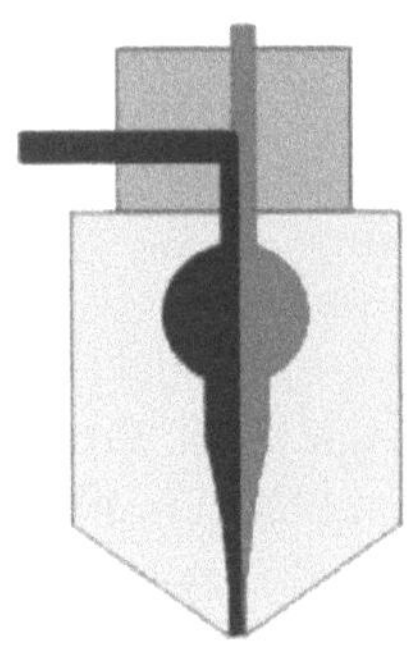

Figure 12.12(a)

Simple combining feedblock

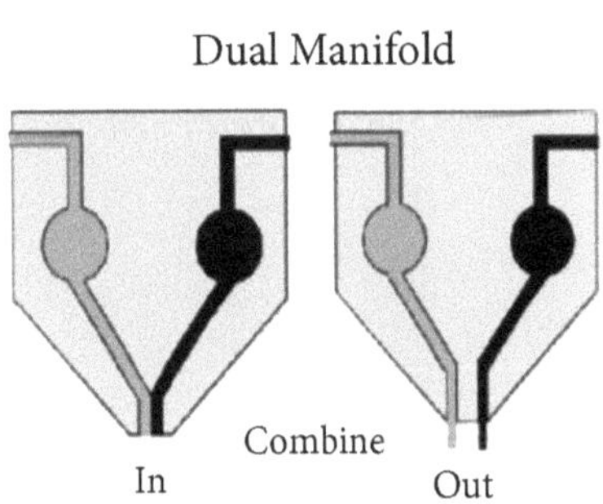

Figure 12.12(b)

Dual manifold options

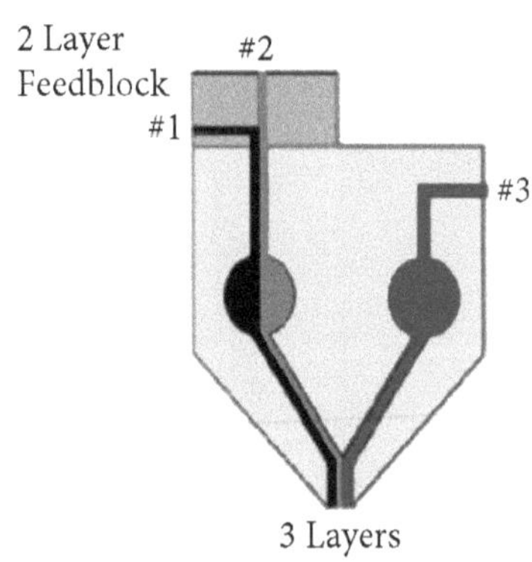

Figure 12.12(c)

Combination die

Figure 12.12

Basic coextrusion layouts for flat dies

The main advantage of the dual manifold system is its greater tolerance to viscosity mismatches as the melt streams are combined in or just outside the die. Dual manifold dies are also more suitable to coextruding layers with large thickness differences. Construction is, however, costly and becomes progressively more complex and unwieldy as the number of layers multiplies. Some other operational comments are:

- Two die adjustments are required, which is difficult.
- Both of the emerging melt streams require sufficient melt strength to avoid rupture or pinholing if combined outside the die.
- This limits the minimum thickness of expensive functional polymers.
- High production downtime is necessary for polymer changes caused by readjustments of each die setting.
- The dual manifold dies are used when it is essential to keep the melt streams at different temperatures.

The role of the feedblock is to bring together the melt streams before they enter the entry port of the die. Feedblocks of various degrees of sophistication are now available some of which can subdivide the melt streams into dozens of microlayers then stacking and recombining them before they enter the die.

12.6 Feedblock vs. multimanifold coextrusion

Table 12.7 summarizes the advantages and disadvantages of the two basic flat die coextrusion systems.

Table 12.7

Feedblock versus multimanifold coextrusion

Attribute	Multimanifold	Feedblock
Application to existing dies	Poor	Good
More than two layers	Poor	Good
Ease of thickness changes	Good	Good
Ability to change layer position	Poor	Good
Skin layer application	Fair	Excellent
Tolerance to viscosity mismatches	Good	Poor
Tolerance for easily degradable polymers	Poor	Excellent
Maintenance of temp. differentials	Good	Poor
Ease of die adjustments	Fair	Good

In **Figure 12.13** a few of the layer combinations that can be produced with a feedblock are outlined. The attributes that can be incorporated in the film are as follows:

- Improved heat-seal by placing an ionomer, copolymer or mLLDPE layer on the outside.
- Incorporation of polyamide, EVOH or polyvinylidene chloride (PVDC) in the structure to provide a high barrier layer.
- Surface properties and appearance can be tailored by placing outer layers of polymers containing additives, pigments etc.
- Buried recyclates in a core layer.
- Degradable or corrosive polymers enveloped in an inert polyolefin to avoid contact with metal surfaces inside the die.

The development of forced encapsulation techniques makes it possible to isolate corrosive and heat-sensitive polymers from direct contact with the internal surfaces of the die. PVDC has been successfully coextruded with die designs of this type. Encapsulation is also a useful method of preventing expensive polymers

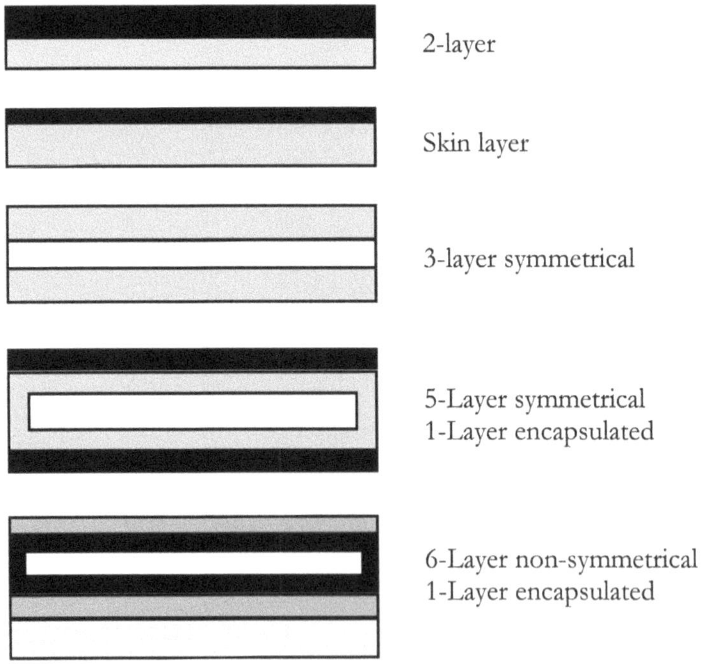

Figure 12.13

Some examples of multilayer structures

from flowing to the outer edges of the die and ending up as edge trim in the waste material. This is essential with polymers such as EVOH. The configurations are only limited by the imagination of the machine designer and processor.

12.7 Flow dividers

Interfacial instabilities can be caused by large differences in velocity between melt streams as they combine inside the feedblock. A method to eliminate this problem is the use of flow dividers. One method of adjusting channel geometry is with moveable vanes or flow dividers inside the feedblock. The externally adjustable vane is rotated about its axis and consequently changes the exit area, i.e. velocity in the flow channel. The intent is to adjust the velocities of each layer so that at the point of intersection the velocities are matched, thus reducing excessive elongation or stretching at the layer interfaces.

The use of flow dividers to adjust velocities at interfaces can be assessed by computer simulations. **Figure 12.14** shows a simple channel where two polymers combine prior to entry in the die. The LDPE is entered at 450 kg/hr and the EAA at 50 kg/hr. The last is 10% of the total mass in the flow domain. The simulated width of the flat die was 1500 mm.

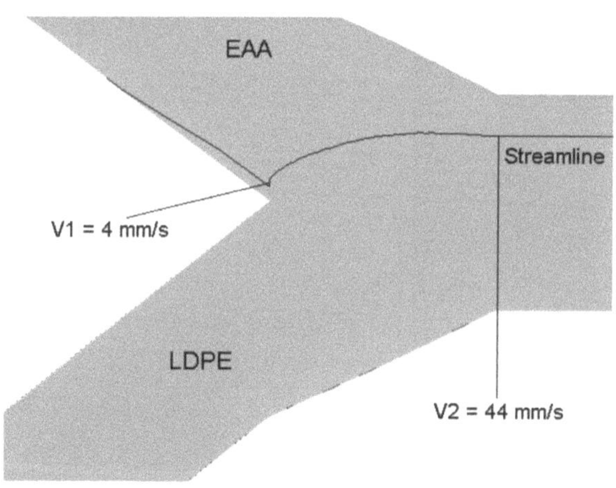

Figure 12.14

Velocity streamline at interface without flow divider

A streamline originating near the EAA wall where the layers combine is also shown. An EAA particle moving along this streamline will accelerate from a velocity of 4 mm/s to 44 mm/s as it combines with the LDPE. This high velocity ratio of ~10:1 may cause wave instability at the interface.

To reduce this velocity difference a flow divider is introduced in the EAA channel. In the simulation this is configured as an internal dam and shown in **Figure 12.15**. The velocity of the EAA particle has now accelerated to 32 mm/s past the divider and remains virtually unchanged at 30 mm/s as it combines with the LDPE and enters the die. The velocity ratio has been reduced to approximately 1:1.

This PC simulation shows how velocities can be controlled by adjusting the geometry of the internal channels to overcome excessive accelerations at layer interfaces. The elongational viscosity of the polymer can play an important role in the outcome particularly with strain hardening polymers such as LDPE and polystyrene where high elongational stresses may be generated.

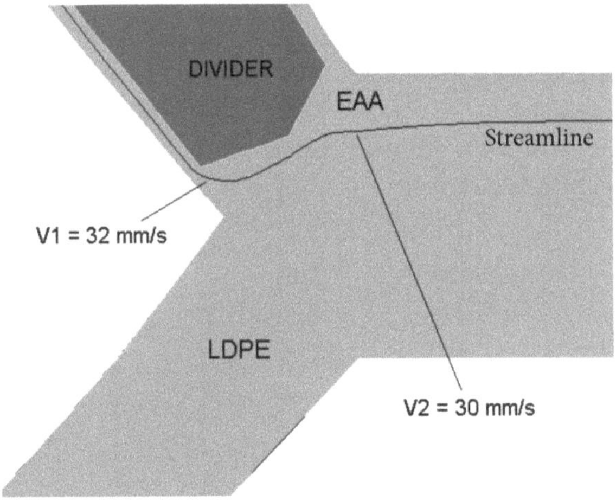

Figure 12.15

New velocity streamline at interface

12.8 Micro-layer coextrusion

This technology is based on splitting coextruded multilayer melt streams and recombining them into micro or nanolayers. As an example in **Figure 12.16**, a symmetrical 3-layer coextrusion is divided into 4 sections, which are then stacked and recombined forming 12 layers, and fed to the multiplier where they are doubled to 24 layers. The skin layers are from the same material as in the original 3-layer film. Thickness is controlled by adjusting the screw speeds of the extruders. The appearance of the film is unchanged, however properties are significantly improved.

The splitting of melts into discrete micro or nanolayers has been shown to improve barrier properties and increase shelf stability of perishable foodstuffs. The microlayers create a more tortuous passage for small molecules. This principal is believed to be successfully exploited in structures containing polyamides and EVOH. Polyamide layers are claimed to be reinforced by some 50% in dart drop and puncture resistance by splitting into multiple layers. In blown film there are claims that if a PA layer is split into two, its burst strength can be increased by 50%. Thermoformability can also be improved with more uniform thinning of the web during forming and yielding greater overall package strength.

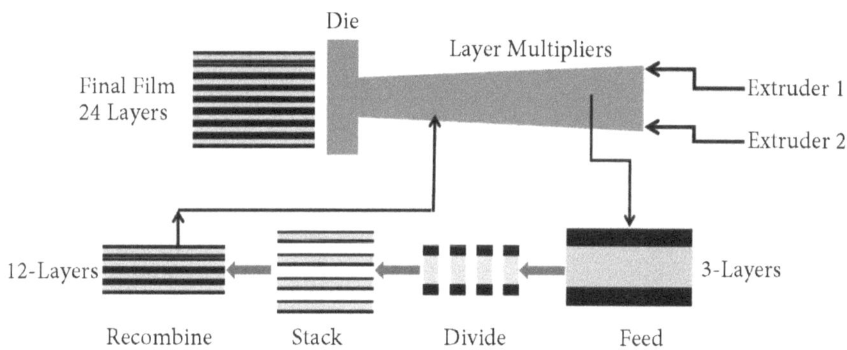

Figure 12.16

Layer multiplier

The physical properties of films such as those used in stretch wrap applications are also claimed to be improved by this process allowing for downgauging by up to 25%.

These developments in coextrusion processes have not only enabled the production of high performance barrier films, but can significantly improve the strength characteristics of simpler polyolefin combinations. Microlayer coextrusions will allow significant downgauging and resource reduction. Since raw materials typically account for some 60% of total manufacturing cost for the processor these technologies offer opportunities to improve competitiveness.

12.9 Coextrusion tie-resins

It is very important for coextruded films to have good interlayer adhesion, which will resist delamination throughout the life cycle of the product. With polymers that do not have any mutual affinity for each other, tie-resins must be placed at the interfaces to bond the layers together. **Table 12.8** lists the types of tie-layer resins used in coextruded structures based on polyethylenes with a range of commonly used polymers.

Table 12.8

Recommended tie-layer resins

Material	Tie-layer resin
PS	EVA, EMA, EnBA, SBS (styrene-butadiene-styrene block copolymers)
PP	Bimodal HDPE, EVA, EMA
PVDC	EVA, EMA, SBS
PA-6	LLDPE-g-MAH (maleic anhydride graft copolymer), EAA, ionomer Zn^{++}
EVOH	PE-g-MAH
PVC	EVA, EMA etc.
PET	Ethylene-butyl acrylate (EBA)-g-MAH, EMA, EVA

It has to be borne in mind that some of these tie-resins may develop different levels of adhesion when used in blown or cast film coextrusion. As a general rule, adhesion in blown film will be higher because of the slower cooling rate of the film, which allows for a longer relaxation and cooling time for the adhesive to interact. In contrast cast film, particularly thin films, are very rapidly quenched. Weak adhesion can also result from interfacial turbulence caused by rapid acceleration as layers combine in the die. This turbulence, such as zigzag waves, may inhibit adhesion between the melt streams. Perfect wetting and contact at the interface is always essential for optimum adhesion.

Chapter 13

Adhesive lamination

The flexible packaging industry requires products that must meet a number of very demanding specifications. When packaging perishable foods no single film will meet all criteria.

Lamination film specifications vary according to the type of goods packed, weight per pack, machinery used, sealing jaws, transportation and shelf stability. This is an important high value added market for PE films. The PE film is primarily used as a heat-seal layer laminated with a polyurethane adhesive or other bonding system to a flexible support as listed below:

- Oriented PET (OPET) film
- Metallized OPET film
- SiO_x (silicon oxide) coated OPET or BOPP films
- Polyamide including OPA (oriented) films
- BOPP film
- Metallized BOPP film
- BOPP plus PVDC or acrylic coating films
- Aluminium foil
- Cellulose films (cellophane)
- Paper

There are some applications where the polyethylene film is extrusion laminated with LDPE to substrates such as paper/aluminium foil combinations. This process requires an extrusion coating machine. These machines are capital intensive but are widely used in applications using paper/board, aluminium foil and LDPE.

Film to film adhesive lamination offers many advantages compared to other combining processes such as extrusion coating and coextrusion:

- Economical for short runs
- Reverse printing is possible

- Excellent print registration
- Good gauge control of individual layers
- Films with excellent machineability
- Combines all roll-fed materials: PE films, oriented and metallized films, aluminium foil and paper
- Combines coextruded films with other supports

In the early days mono films of LDPE and EVAs (4-5% VA) were the standard for lamination applications. The development of LLDPE and advanced metallocene polyolefins now offers the processor a much wider choice of options particularly an infinite number of blended formulations. Add to this the many coextrusion possibilities and it becomes impossible to recommend the best formulation for each application. The recommendations discussed here are mostly suggestions rather than specific formulations.

The lamination films must be corona treated to 42-45 mN/m (see chapter 15 on corona treatment). On blown film lines thickness variation should be kept below 8% (2σ) of the target value. For some applications it may be necessary to reduce the variation to 5%. These tight gauge requirements can only be met with good die designs and cooling rings plus automatic thickness control systems. Weak seals and leakers can very often be traced to excessive thickness variations. Oscillating dies or rotating haul-offs are necessary. The collapsing and winding systems must deliver perfectly flat rolls wound at uniform tension. Centre or gap winders are preferred. It may also be necessary to edge trim the layflat (~20 mm) so that the finished rolls have perfectly aligned edges. If this is done the trim should be recycled in-line to minimize scrap losses. Devices that collect, compact and chop up the trim must be included in the design of the blown film equipment. The chopped up trim is recycled via a screw transport feeder back into the extruder. Since most lines used today for high performance lamination films incorporate coextrusion the trim is best recycled in the core layer of an ABA or ABC structure.

Poor interlayer adhesion is a continuing source of product defects. Many are caused by excessive slip additive, faulty formulation and inadequate adhesive curing. Slip agent concentration must be rigorously controlled. Too much slip will weaken adhesion and/or affect heat-seal strength. Not enough slip will lead to a high COF and poor machineability on the downstream packaging equipment. Erucamide, which migrates more slowly than oleamide is the preferred slip additive in lamination films.

In coextrusion, many processors use polyethylenes that contain slip and antiblock in all layers. Antiblock (silica) in the inner layers is wasted. Slip agents such as erucamide will always find their way to the film surfaces. In a 3-layer structure, the slip agent in all layers will migrate to the surface. If each layer contains 500 ppm erucamide the total will effectively add up to 1500 ppm all of which will end up on the outer skins. This may be excessive and cause delamination. A better option is to place the slip agent in the core layer only. This will migrate outwards and within a few hours will yield the equivalent of the original 500 ppm on the surfaces of the 3-layer film.

The addition of 20-30% LLDPE to LDPE or EVA will improve hot tack and tensile strength. For the highest hot tack and increased heat resistance (e.g. boil-in-bag applications), the LLDPE addition should be increased to 80% or mMDPE used. In some cases HDPE is blended with the lower density resins. It is important to note that with linear rich blends the bubble stability will deteriorate, thus making it more difficult to minimize film thickness variations. Dual lip cooling rings should be used with high LLDPE concentrations.

The mPEs (metallocenes) are excellent candidates for lamination films. Adding 15-20% metallocene PE (POE) to LDPE will significantly improve hot tack, reduce SIT and lower haze. These very low density metallocene resins have low extractables, unlike Z-N PEs, and comply with food packaging legislation. Haze and dart impact strength values comparing EVA (9% VA) and mPE are shown in **Table 13.1**. In applications requiring densities below 0.915 g/cm^3 i.e. low SIT, it is recommended to use metallocene PEs.

Table 13.1

Haze and impact strength comparisons

Property	Unit	mPE	EVA (9% VA)
Haze	%	3	3
Dart impact	g/mil	>1050	200

The interest in metallocene polyethylene for the laminator is principally in their outstanding sealing performance. However second generation resins made at higher density could also contribute some advantages. These relatively stiff materials with good transparency are well worth further investigation in blends or coextrusions by laminators.

The development of polypropylene random copolymers (PPRC) has enabled the production of very high clarity films using the cast film process (CPP). These have good sealing properties and can be used in applications were improved heat and oil resistance is specified. CPP seal layers can resist retorting conditions.

13.1 The lamination process

In adhesive lamination a number of considerations must be taken into account in the selection of the bonding system:

- Shelf-life of reagents
- Pot-life after mixing
- Nip pressure and temperature
- Curing time and temperature
- Mixing ratio of components
- Solution viscosity
- Drying temperature and air velocity
- Initial bond strength (green tack)
- Final bond strength
- Product resistance

Figure 13.1 shows the layout of a basic film lamination machine. The adhesive solution is spread from a gravure roll applicator onto the film (primary web). The gravure roll is usually a steel roll or cylinder engraved with cells that pick up the liquid, which is applied on the web and passed through a nip. The application roll will usually have a surface finish of 40-70 lines/cm. Typical solvents are ethyl acetate, methyl ethyl ketone (MEK) and alcohols. The adhesive should have a surface tension at least 10 mN/m lower than the surface being coated.

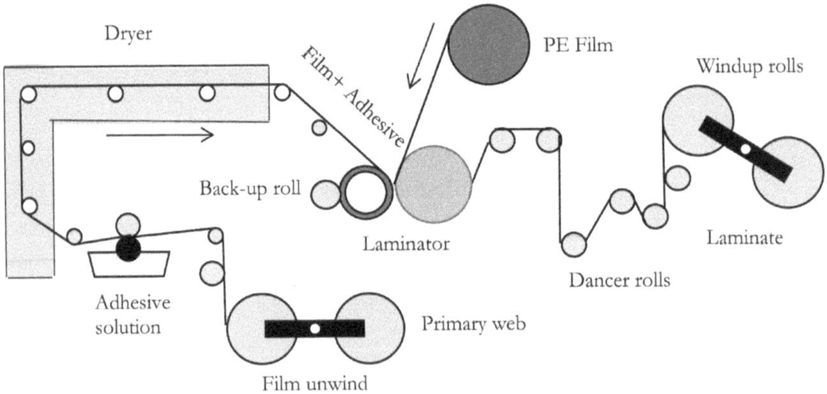

Figure 13.1

Adhesive laminator

The dry coating weight, which can be as low as 1 g/m^2, will depend on the applicator configuration and the viscosity of the adhesive solution. Drying is carried out in a heated tunnel with high velocity hot air. The dried film is then laminated to the polyethylene film in a heated nip (40-70 °C) as shown in **Figure 13.1**. A steel back-up roll to prevent deflection is normally included. To avoid film curling, the two films entering the nip must be at equal tension. DC drives capable of accurate tension control must be used. It is important that the adhesive layer is not rubbed by contact with rolls or other surfaces prior to entering the laminating section.

It is recommended to set the tension at the winder higher than at the unwinder. The tension should be progressively increased as the web flows downstream. Some typical adhesive applicator configurations are shown in **Figure 13.2**.

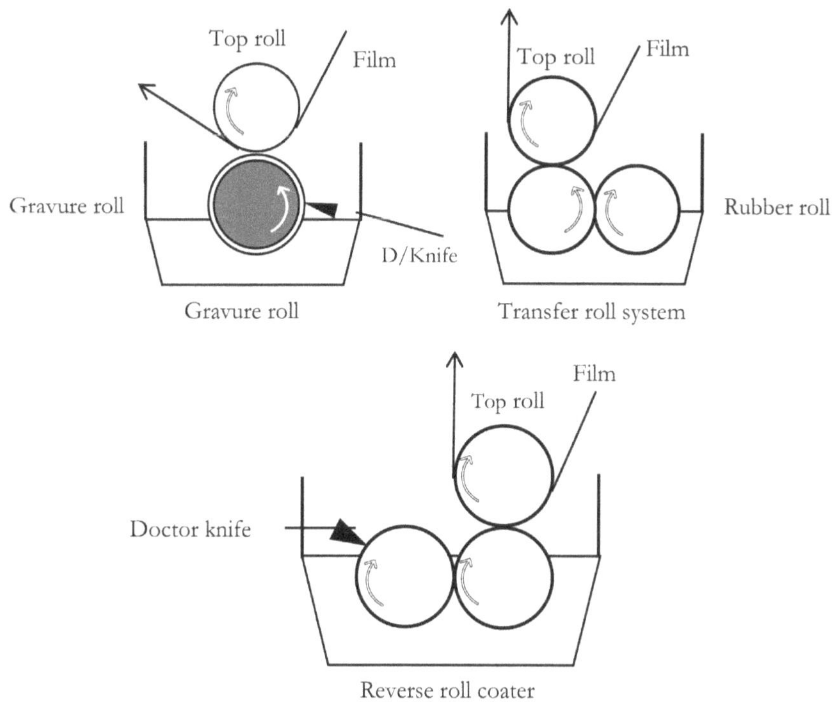

Figure 13.2

Typical adhesive applicators

Since the adhesive will generally have an initial low tack, it is important that tension control is precise to avoid any delamination between the laminating nip and the winder. Dancer rolls should be fitted as required to facilitate web handling and avoid creasing and wrinkles. These are either pneumatically or electronically operated, to equalize web tension in both the machine and transverse directions Ageing for a week is normally required to complete the cure of isocyanate-based adhesives.

To avoid roll deflection typical roll specifications and dimensions as shown below are normally used. Nip pressure should be in the range of 10-50 kg/cm width. The nip rolls should be perfectly parallel with uniform pressure distribution across their width.

Top roller	Durometer hardness Shore A: 35-40 Sleeve type: neoprene, silicone or polyurethane
Bottom roller	Nickel or chrome plated steel or hard rubber
Typical roll width (cm)	75-80
Roll diameter (cm)	15-20

Some of the following problems can occur during the lamination of two films:

Defect	**Adjustment required**
Wrinkles, poor bond, web curl	Tension control
Internal laminate stresses will cause web curl	In-out feed tension control
Weak bond, adhesive ooze, wrinkles	Nip adjustment, tension control
Air bubbles, wrinkles	Contact point, poorly aligned rolls

Figure 13.3 shows possible defects that can occur with poorly machined or maladjusted nip rolls. Adhesion will be uneven across the web from concave, convex and canted shaped rolls. Other defects are curling, bubbles and wrinkle formation. At the nip, the rolls must be perfectly flat with uniform pressure across the width.

Nip pressures can be controlled manually or pneumatically with air cylinders. There should be a single control for rapid and uniform adjustment of the actuators. Non-uniform pressure will also lead to poor web steering in addition to the above defects. The polyethylene film must be brought in contact with the adhesive layer at the nip contact point and not earlier otherwise air entrapment can occur.

Figure 13.4 shows schematically a lamination combining reverse-printed OPET with LDPE film. It is important to ensure that there are no interactions between the adhesive, the printing inks and additives in the film.

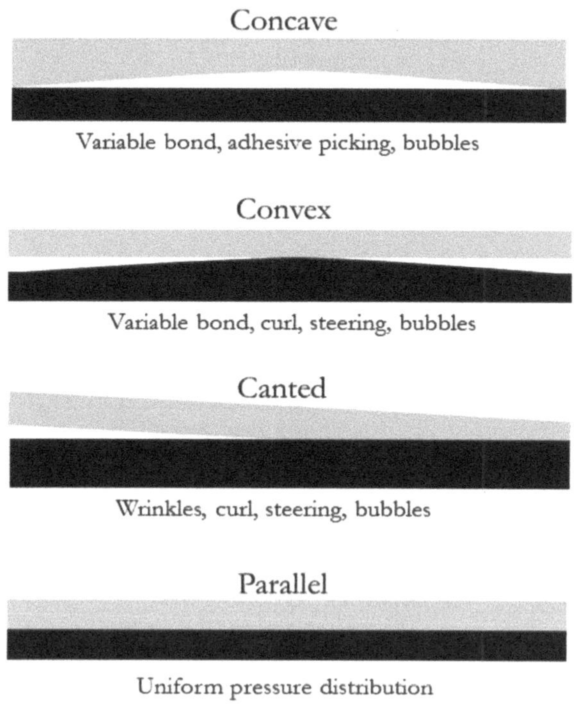

Figure 13.3

Interfaces between faulty applicator rolls

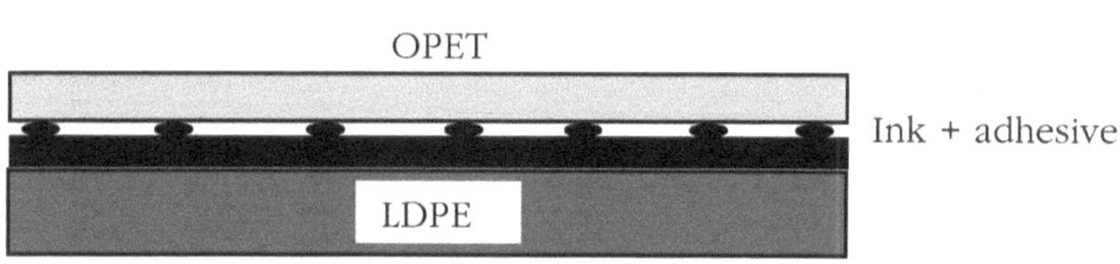

Figure 13.4

Typical laminated film structure

13.2 Film laminating adhesives

The adhesives used in high performance film lamination are available in the following categories.

- Solvent-based single component

- Solvent-based two components
- Solventless single and two component
- Aqueous or water borne

A snapshot comparison of the above systems is shown in **Table 13.2**.

Table 13.2

Subjective comparison of basic adhesive systems

Solvent-based	Aqueous	Solventless
Urethane-based	Urethane-based	Urethane-based
Solvent emissions	No organic solvent	No emissions
Medium adhesive cost	Medium adhesive cost	Low adhesive cost
High energy needs	High energy needs	Low energy needs
Medium line speed	Medium line speed	High line speed
Low viscosity	Low viscosity	Medium viscosity
Curing during lamination	Curing after lamination	Curing after lamination

The most commonly used adhesives are those based on polyether or polyester-urethane chemistry. A urethane is formed by reacting an isocyanate with a hydroxyl containing reagent, e.g. polyol.

R-NCO + HO-R′ → R-NH-CO-O-R′

Isocyanate + Polyol   Urethane

A polyurethane is synthesized by reacting a di-isocyanate, typically toluene di-isocyanate (TDI) or 4,4-diphenylmethane di-isocyanate (MDI), with a diol as follows. The use of TDI has been reducing owing to health concerns.

OCN-R-NCO + HO-R′-OH → (-CO-NH-R-NH-CO-O-R′-O-)

Diisocyanate   Diol   Linear polyurethane

The chain is extended by reacting the di-isocyanate with hydroxyl terminated polyesters or polyethers. These are available as single or two component adhesives. The polyether polyols are usually more heat resistant than their polyester counterparts.

13.2.1 Single component adhesives

A single component, moisture curable adhesive forms a modified polyurea. Carbon dioxide CO_2 is released.

$$R\text{-}N{=}C{=}O + H\text{-}O\text{-}H \rightarrow (R\text{-}\underset{\displaystyle H}{\underset{|}{N}}\text{-}\underset{\displaystyle O}{\underset{\|}{C}}\text{-}OH) \rightarrow R\text{-}NH_2 + CO_2$$

The generic primary amine will instantly react with another -NCO group, which will initiate chain propagation.

$$R\text{-}NH_2 + O{=}C{=}N\text{-}R' \rightarrow R\text{-}\underset{\displaystyle H}{\underset{|}{N}}\text{-}\underset{\displaystyle O}{\underset{\|}{C}}\text{-}\underset{\displaystyle H}{\underset{|}{N}}\text{-}R'$$

The release of carbon dioxide may cause blistering with non-porous films. When laminating to hygroscopic polymers, such as polyamide, residual moisture in the film may react with the isocyanate releasing CO_2. This may also cause blistering in the laminate.

13.2.2 Two component adhesives

Two component adhesives are produced by mixing a prepolymer with free isocyanate groups with a resin containing free hydroxyl groups (polyol). The two components will polymerize (cure) to form an adhesive film. Excess isocyanate will react with moisture in the atmosphere and/or substrate.

$$R\text{-}N{=}C{=}O + HO\text{-}R' \rightarrow R\text{-}\underset{\displaystyle H}{\underset{|}{N}}\text{-}\underset{\displaystyle O}{\underset{\|}{C}}\text{-}O\text{-}R'$$

However, if there is excess alcohol, e.g. ethanol or isopropanol present from inadequate drying of printing inks the adhesive will be terminated by the alcohol. This will form an adhesive layer which is tacky and will have weak heat resistance. The ethanol will terminate the polyurethane, which will lead to a low molecular weight chain.

$$R\text{-}N{=}C{=}O + HOC_2H_5 \rightarrow R\text{-}\underset{\displaystyle H}{\underset{|}{N}}\text{-}\underset{\displaystyle O}{\underset{\|}{C}}\text{-}O\text{-}C_2H_5$$

The isocyanate component will usually have a viscosity of 1.6-3.0 Pa.s and the resin component a viscosity of 5.0-15.0 Pa.s. It is essential that the cure is complete otherwise there is the possibility of the monomers migrating to the surface with negative effects, such as weak heat seals.

In order that the adhesive be effective, a threshold molecular weight must be achieved. With low reactivity polyolefins films high molecular weights are required to achieve useful properties. Typical molecular weights of urethane adhesive systems are shown in **Table 13.3**.

Table 13.3
Molecular weights of urethane adhesives

System	M_w	M_n
Solvent-based	12,000	6,000
single component solventless	9,000	5,000
Two component solventless	6,000	4,000
Aqueous	72,000	40,000

The aqueous adhesives are dispersions and can, therefore, be polymerized to high molecular weights as their viscosity in water will be low enough to be applied from a roller coater. In contrast the solventless systems are prepolymerized to low molecular weights in order to keep their viscosity sufficiently tractable for application. There is no solvent present to act as a diluent and wetting agent.

The components must be well mixed and in the correct proportions and evenly spread on the film surface. The ratio of isocyanate to hydroxyl and the laminating and storage conditions will influence the quality of the bond. The lower the viscosity the easier it is to spread the adhesive evenly and to achieve a uniform dry weight on the surface. The adhesive bond is made between active radicals on the film's surface and the reactive component from the adhesive. In the case of polyethylene a corona treated surface with a surface tension of 42-45 mN/m is normally required for interaction with the adhesive. OPET films have a high surface energy and are the easiest to laminate. BOPP films are the most difficult and require intensive corona or flame treatment.

The key requirements of laminating adhesives are:

- Non toxicity
- Good green tack
- Fast solvent release
- Film forming at low coating weight
- Good wetting
- Long pot life
- Absence of volatile byproducts

In the finished laminate the adhesive must fulfil the following criteria:

- Permanent bonding
- Heat stability

- Elasticity retention
- Colour stability
- Compliance with food packaging regulations

With the drive towards reduced environmental pollution from solvents, attention has focused on solventless formulations. Because of their higher viscosity these systems are more difficult to apply uniformly at the desired low coating weights of 1-2 g/m^2. Solvents will also wet the film surface and wash away contaminants thus enhancing interface contact. This added benefit is not applicable with solventless formulations.

13.2.3 Solvent-free PU adhesives

Environmental pressures are driving converters to reduce their use of organic solvents. Adhesive suppliers have reacted by offering solventless laminating adhesives that cover a wide range of films and foils with good product resistance. One advantage of solventless systems is that they do not require a drying oven, which saves space and energy. Application weights are 1.5-4.5 g/m^2.

As outlined in the preceding section solvent-based adhesives are prepared in two steps. A PU prepolymer is first produced and then diluted with a solvent prior to the molecular weight increasing to an unmanageable level by chain propagation. The diluent keeps the overall viscosity low for easy application on the substrate. The molecular weight is controlled prior to application and then cured in the final laminate.

In the solventless system the reaction in the first step is stopped earlier. The prepolymer is now of lower molecular weight and of low enough viscosity to be applied on a gravure or similar coater. Curing takes place at ambient temperature although some converters may condition the laminates in hot rooms for several days. Since it is applied at a lower molecular weight more polymerization takes place in the solventless system after application in the laminator.

Problems may arise with single component isocyanate-terminated systems since these are cured with moisture. A moisture spray is usually installed on the machine. The risk here is that reaction may occur with retained alcohol in printing inks rather than moisture, leading to an uneven cure. However, with care these solventless single component adhesives can be used with a variety of substrates so long as these are lightly printed. Generally polyether-based urethane adhesives are preferred for film and paper substrates and work well with highly printed surfaces. For aluminium foil laminations polyester systems should be used.

In a two component formulation where curing takes place between the isocyanate and the hydroxyls in the curing agent this problem is less pronounced. Solventless two component adhesives are available in two forms:

- Low viscosity applied at ambient temperature
- High viscosity applied at 40-70 °C

High viscosity adhesives have the advantage of developing higher initial bond strengths. The low viscosity systems are used with heat-sensitive films, but because of their low initial molecular weight they will have lower initial bond strength.

The polyether systems, although they wet better and develop strong bonds, tend to absorb migratory molecules such as the slip additives and waxes usually present in PE. This may increase the COF of the laminates. The polyesters on the other hand may develop weaker bonds, but do not absorb slip additives and, therefore, have a minimum effect on the laminate's COF. However, if the slip agent concentration is too high bond strength will drop. Combining the two functional groups into one system is one approach to optimizing the performance of the laminate. Reducing the coating weight of the adhesive can also help to minimize the migration of small molecules.

Adhesive suppliers have developed solventless formulations that combine polyether and polyester groups, which optimize surface wetting and have a minimal effect on slip agent loss. These adhesives are recommended for high speed form, fill and seal (FFS) applications.

Figure 13.5 is a schematic layout of a solventless adhesive applicator. The primary web is coated with the adhesive and laminated to the polyethylene film (secondary web) in the laminator nip as shown. No drying oven is required. In the absence of

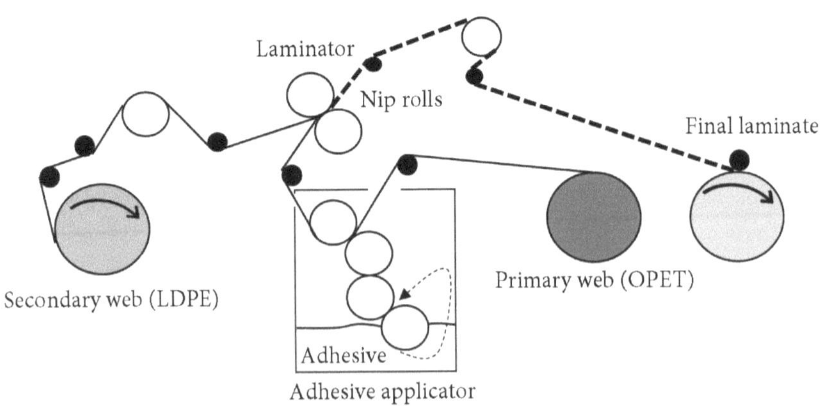

Figure 13.5

Schematic of solventless adhesive laminator

a solvent, which helps to wet the film surface, the PE film must be corona treated to 42-45 mN/m to ensure that the adhesive is well anchored to its surface. The preferred applicator is by smooth rather than gravure roll.

In **Table 13.4** the data shows the dependence of the laminate's coefficient of friction with slip agent (erucamide) concentration in the PE film. The aluminium foil/LDPE laminates were produced with two different two-component solventless adhesives. The results show that adhesive #2 develops a COF which is too high with erucamide levels below 500 ppm after 1 month ageing. Adhesive #1 appears not to be sensitive to the slip agent concentration.

In **Table 13.5** some adhesion and heat-seal strength data for a number of laminated structures are shown. A solventless two-component polyether-polyester adhesive was used in all cases.

Table 13.4
Effect of slip agent concentration on COF

Al Foil/LDPE	1 Week	1 Month
	150 ppm erucamide	
Adhesive #1	0.21	0.21
Adhesive #2	0.21	0.42
	300 ppm erucamide	
Adhesive #1	0.2	0.19
Adhesive #2	0.2	0.32
	500 ppm erucamide	
Adhesive #1	0.21	0.19
Adhesive #2	0.19	0.18

Table 13.5
Adhesion and seal strength of some typical laminates

Laminate	Bond strength (8 days) (N/15 mm)		Seal strength (N/15 mm)
	Printed	**Unprinted**	
OPA 15 μm/EVA 50 μm	-	4.5	34
PA 30 μm/EVA 50 μm	-	4.7	35
OPA 19 μm/EVA 50 μm	3.0	Inseparable	33
OPET 12 μm/LLDPE 50 μm	3.1	Inseparable	45
OPET 12 μm/Al foil 12 μm	2.5	2.9	-
Met. OPP 19 μm/OPP 19 μm	1.9	2.6	-
Met. OPET 12 μm/LLDPE 60 μm	-	3.2	41
Courtesy Morton Adhesives			

Product resistance data for a two-component solventless adhesive are given in **Tables 13.6** and **13.7**. This adhesive can be used with a wide range of film and foil combinations and can be classed as a medium performance adhesive comparable to solvent-based systems.

The following laminate was produced. Adhesive coating weight was 2.5-3.0 g/m^2. The LDPE contained 350 ppm erucamide.

OPET (12 μm)
Al. foil (8 μm)
LLDPE (60 μm)

These results show poor product resistance to acids and alcohol. Aluminium foil is attacked by acids, which explains the poor results in acidic conditions. Adhesion values to curry powder, paprika, mayonnaise and detergent were marginal.

In **Table 13.7** similar data is generated but this time with the following laminate.

OPET (12 μm)
LDPE (70 μm)

Resistance to acids is much improved by replacing the foil with polyester film. However, resistance to alcohols is still poor.

Table 13.6

Product resistance of solventless adhesive, OPET/Al/LLDPE laminate

Product	1 Week @ 45 °C	1 Month @ 45 °C
	(N/15 mm width @ 100 mm/min)	
Air	7.0-11.0	10
Water	4.8-6.3	6.5-8.0
Curry powder	2.0-2.5	1.5-2.5
Paprika powder	3.0-5.0	2.5-3.5
Shampoo	4.0-5.5	4.0-5.5
Liquid detergent	2.3-3.0	2.3-3.0
Mayonnaise	2.5-3.3	2.3-3.0
Capri-sun fruit drink	4.0-5.5	3.5-4.8
Acetic acid (1%)	0.5	0.4
Isopropanol (90%)	0.9-1.3	1.2-1.5
Ethanol (99%)	0.8-1.0	1.0-1.1
Motor oil	5.8-8.0	5.7-8.0
Storage: flat		

Although there are pressures encouraging the use of solventless adhesives it is important that they have the minimum residues of free isocyanate or low molecular weight species in order to both meet food compliance legislation and minimize the risk of surface contamination, which might compromise seal integrity.

Table 13.7

Product resistance of solventless adhesive, OPET/LDPE laminate

Product	**1 Week @ 45 °C**	**1 Month @ 45 °C**
	(N/15 mm width @ 100 mm/min)	
Air	N. S.	N. S.
Water	N. S.	N. S.
Curry powder	2.5 tear	N. S.
Paprika powder	2.1 tear	2.8 tear
Shampoo	N. S.	N. S.
Liquid detergent	N. S.	N. S.
Mayonnaise	N. S.	N. S.
Capri-sun fruit drink	N. S.	N. S.
Acetic acid (1%)	N. S.	N. S.
Isopropanol (90%)	0.9-1.5	0.6-1.5
Ethanol (99%)	0.1	0.1
Motor oil	N. S.	N. S.
Storage: flat		
N. S. = not separable		

13.2.4 Water-based adhesives

Water-based adhesives are aqueous polyurethane or acrylic dispersions. These are fully reacted polymers characterized by high molecular weight and are delivered ready to use to the laminator. The high MWs reduce the risk of migration of unreacted monomer. The solids content in the dispersion is 44-48% and they are usually available as one-component systems. Thermal resistance can be improved by reacting with aliphatic crosslinking agents. These do not contain aromatic isocyanates. The adhesion process is not based on chemical reactions of the reagents but physical interaction (wetting) at the interfaces. Application is usually by gravure coating heads. Line speeds up to 800 mpm are claimed. Corona treatment should be at a higher level than normally used with solvent-based adhesives. A recommended target for surface tension is 44-45 mN/m.

These adhesives can develop high peel strength immediately after lamination in contrast to the conventional isocyanate formulations. In-line converting from lamination, printing and slitting is possible. This allows for much faster deliveries.

Some typical experimental bond strengths with water-based polyurethane adhesives are shown in **Table 13.8**.

Table 13.8

Bond strengths of typical laminates

Laminate	Initial bond (cN)	3 Days (cN)	1 Month (cN)
OPP/PE	120-260	150-180	260-330
PA/PE	200-250	500-700	400-650
BOPP/m-BOPP	110-130	110-140	200-280
PET/PE	200-300	460	420
m-PET/PE	110-130	170	170
Al foil/PE	160-190	220-240	260-280
PET/Al foil	250-350	350-420	230-280

13.3 Some precautions regarding laminates

Laminates are heterogeneous structures which embody different films, adhesives and inks. All or some components may contain migratory products such as slip agents, waxes, residual solvents and plasticizers. These materials may produce odours or off-taste and affect the seal or bond strength and machineability on the packaging lines. It is therefore important that all materials are carefully selected to avoid any long term defects appearing in the finished package. For instance with certain adhesive formulations the temperatures at the lamination nip and storage can affect the final COF of the film. Polyether-based adhesive systems are believed to be more prone to this effect.

Figure 13.6 shows a combination printing and lamination line to produce a typical BOPP-based snack food structure.

Printing ink formulations should be free of aromatic solvents (toluene) and lead chromate pigments. Other undesirable materials are chlorine-containing binders, phthalate and diarylide pigments.

The adhesive must completely wet the film surfaces including the printed areas. The inks must be completely dried and well adhered to the surfaces. There should be no interaction between the adhesive and the ink formulation. As discussed elsewhere, the polyethylene film must be correctly formulated with slip and antiblock additives. Surface treatments such as corona or flame treatment must be optimized for all film surfaces. Flame treatment is preferred by some laminators to avoid the risk of back treatment and pinhole formation on very thin films.

Excess slip additives will weaken the bond and seal strength. The reverse will lead to poor performance on form, fill and seal machines. Acid or oily foods may attack the adhesive causing delamination. Therefore, product resistance data must be generated for the proposed application. Some acid foods (pH<6.4) may attack aluminium foil causing failure of the laminate. In these circumstances alternate barrier materials such as PVDC, EVOH and/or polyamide should be considered.

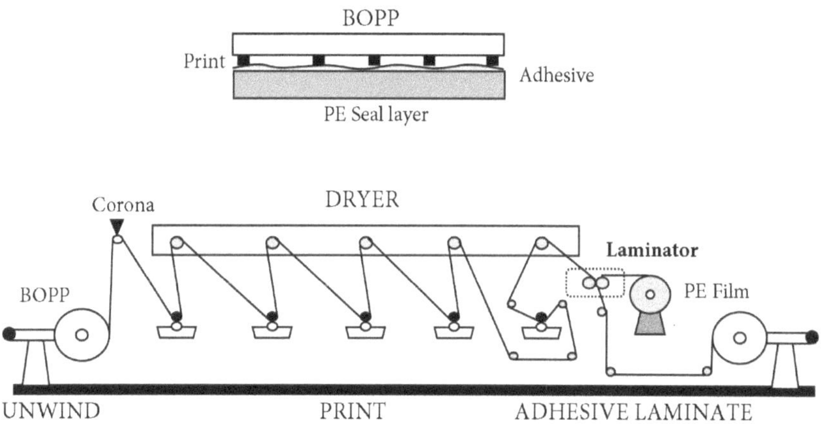

Figure 13.6

In-line printing and lamination

Heat resistance of the structure must be verified if the packed product is to be pasteurized or sterilized.

13.4 Organoleptic properties

It is important to be aware of the complex interactions that may occur between a food product and the laminate. Any migration effects, i.e. scalping of flavours and nutrients into the film will result in a loss of quality and should be taken into account when selecting materials and film thickness, particularly the polyolefin contact layer. Many aroma compounds and nutrients present in natural foods can be transported into the amorphous phase of all polyolefins.

This scalping phenomenon is linked to the composition and morphology of the polymer, which determines its permeability (P). Permeability is a function of the solubility parameter (δ) and diffusion rate constant (D). The basic relationship is given in equation 13.1.

$$P = D \times \delta \tag{13.1}$$

The absorption interaction constitutes the migration of specific organic compounds from the packed product into the polymer forming the contact layer.

The respective solubility parameters (δ) give a good indication of the mutual affinity of a polymer for a specified substance. The solubility parameter of a polymer according to van Krevelen[1] is defined as the square root of the cohesive density in the amorphous state at room temperature. The constant is expressed as:

$$\delta = \sqrt{\frac{E_{coh}}{V}} \tag{13.2}$$

E_{coh} is the cohesive energy of the substance and V is its volume per mol. The solubility parameter (δ) is expressed in J/cm³. **Table 13.9** shows the varying degrees of absorption (scalping) for a range of plastic films in contact with a variety of organic compounds.

Table 13.9

Scalping of polar and non-polar polymers

Compound		PE	PP	PVC	PET	PA (d/w)	PVDC	EVOH (d/w)
	δ (J/cm³)	15.6	16	19.4	21.4	30	24.4	32
% Loss or scalping after conditioning								
Hexane	14	80	65	5	2	<1/3	<1	<1/1
Octane	16	75	60	10	2	1/3	<1	<1/2
Cyclohexane	16	70	60	2	1	1/2	1	1/3
D-limonene	18	70	65	15	4	1.5/5	<1	1/3
Citral	18	50	50	30	12	1/5	1	<1/5
Menthone	18	65	50	50	3	2/9	2	1/5
Et. acetate	18	35	30	45	8	4/5	1	2/10
Carvone	18	50	50	40	4	3/7	1	1/7
Anethole	20	35	30	75	25	1/3	<1	<1/7
Menthol	20	40	30	25	2	3/3	2	2/12
Benzaldehyde	20	45	40	60	15	4/11	2	2/15
Octanol	20	35	20	25	5	2/6	<1	<1/3
Me salicylate	22	40	35	20	7	5/7	1	3/12
d/w:dry/wet.								

[1] van Krevelen, D.W. and te Nijenhuis, K., Properties of Polymers, 4th Edition, 2009, Elsevier Science, 978 0080548197, Ch. 7

Scalping by the two polyolefins is far higher than the polar polymers such as EVOH and polyamide. The polyolefins have solubility values of 15-16 J/cm^3 compared to 30-32 J/cm^3 for the polar materials. The δ values for the organic compounds are in the range of 14-20 J/cm^3 and closely matched to the polyolefins and hence more readily absorbed. Losses in flavour and nutrients can be reduced by artificially "spiking" the product to be packed.

13.5 The anti-sealing effect in laminates

It is important that a laminating adhesive is completely cured and none of the monomers are free to migrate to the polyethylene surface as this will compromise the heat-seal strength of the film. This anti-sealing effect was first reported in the late 1970s at a time when laminators were trying to replace toluene di-isocyanate (TDI) with 4,4-diphenylmethane di-isocyanate (MDI) in adhesive formulations. This development was spurred by the tightening of legislation regarding the use of suspected carcinogens in the work area. MDI was preferred because of its lower vapour pressure at ambient temperatures. However, adhesive systems based on MDI were believed to contain a higher level of extractables increasing the risk of migration of unwanted substances at the film surface, leading to undesirable after-effects such as resistance to heat-sealing (anti-sealing).

Anti-sealing manifests itself as loss in seal strength of the laminate at the normally accepted seal temperature, time and pressure. The phenomenon occurs sporadically making it difficult to isolate its origins. In the following example analysis of the non-sealing film surface has shown the presence of polyurea. Sealability is usually restored after washing the surface with a solvent, such as isopropanol, and the presence of MDI monomer in the adhesive was believed to be the culprit. Unreacted MDI migrates to the surface of the polyethylene seal layer, reacts with moisture and forms the amine by splitting off carbon dioxide (CO_2). The amine then reacts with residual di-isocyanate to form urea and polyurea.

The generic reaction is shown below.

$$\underset{\text{di-isocyanate}}{OCN - R - NCO} + 2H_2O \rightarrow \underset{\text{amine}}{H_2N - R - NH_2} \xrightarrow{+OCN\text{-}R\text{-}NCO} \underset{\text{polyurea}}{(-NH-R-NH-\underset{\underset{O}{\|}}{C}-NH-R-NH-\underset{\underset{O}{\|}}{C}-)}$$

The molecular weight of the polyurea at the surface will determine the resistance to sealing at a given temperature. **Figure 13.7** compares the effect on seal strength of the MDI with and without the curing agent. The drop in seal strength is catastrophic.

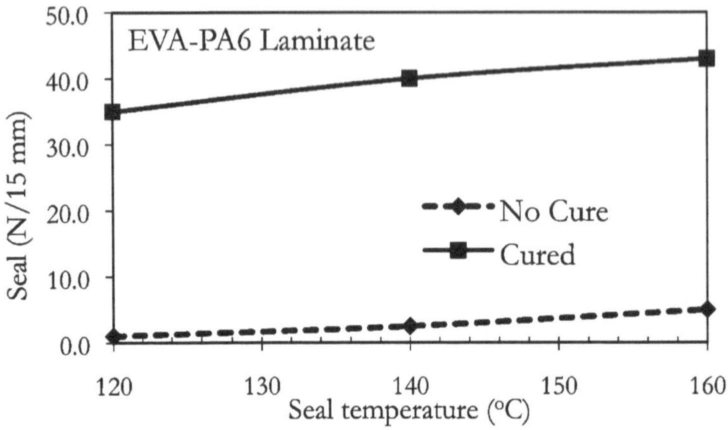

Figure 13.7

Effect of curing agent on seal strength
Adhesive: 2 g/m²

Adhesive suppliers are continually developing faster curing formulations that do not retain any free isocyanate that may contaminate the surface of the laminate. However, complaints about the loss of seal strength are still commonplace in the industry.

It must also be borne in mind that slip (erucamide) concentration may also have an effect on seal strength. **Figure 13.8** shows the effect of slip agent concentration on the seal strength of an EVA (5 VA%) film laminated with an MDI solventless

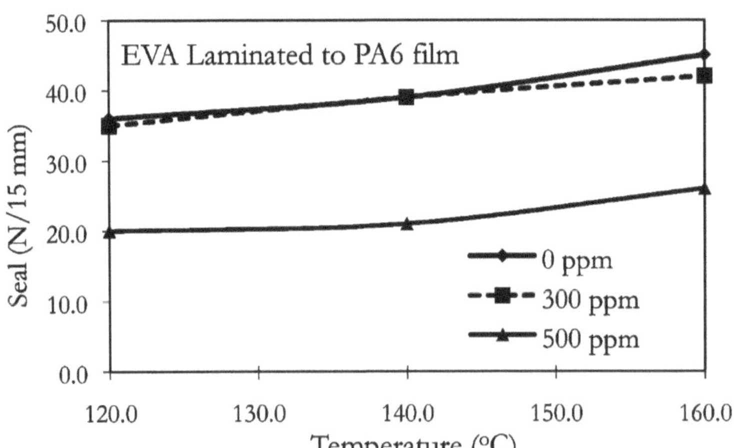

Figure 13.8

Effect of slip agent level on seal strength

adhesive to a 50 μm polyamide film. The EVA film was 100 μm in thickness. This data shows that the presence of slip at concentrations above 300 ppm may also create an anti-sealing effect at this film thickness. Slip concentration should be adjusted not only according to the COF required, but also according to thickness. Thick films will contain more available slip to migrate to the surface. In this example with a 100 μm film, erucamide concentration >300 ppm is clearly excessive.

Chapter 14

PE Formulations for film applications

A number of additives are formulated with polyethylenes in order to modify the surface characteristics of the film to facilitate handling and converting. Many of these additives are included by the polymer supplier, however for more specialized applications the modification is carried with specialized masterbatches added by the processor. It is important to produce films that have the optimum surface friction and layflat tubing and bags that are easily opened and free of static.

14.1 Slip additives

The COF (coefficient of friction) is a measure of the ability of a surface to slide over an adjoining surface. The force to pull the moving sample is measured according to standardized procedures and usually reported as two levels of COF:

1. **Static COF**: The force required to initiate movement of the surfaces relative to each other.
2. **Kinetic COF**: The force required to sustain this movement.

Slip is a qualitative term describing the lubricity of the surfaces sliding over each other. A high slip value corresponds to a low COF and vice versa.

The following fatty acid amides are the most commonly used as slip additives:

Oleamide	C_{18} primary amide
Erucamide	C_{22} primary amide
Unsaturated	C_{36} secondary amide
Unsaturated	C_{40} secondary amide

COF is adjusted by the addition of slip agents to the polymer. The slip agents work on the principle of their incompatibility with the host polymer, which encourages migration to the film's surface where they act as lubricants. **Figure 14.1** illustrates the migration mechanism as the slip agent exudes from the molten polymer towards its surface on solidification. Over a short time span the molecules are rearranged on the surface to create a low friction layer. The amide groups are in contact with the surface and the hydrocarbon chains are projected outwards to provide the low friction surface.

Migration rate is inversely proportional to the molecular weight of the slip agent. These fatty acid amides are natural products and their purity may vary considerably from different suppliers. Slip formulations will range from 100-1000 ppm for LD/LLDPE. A typical medium slip concentration is 500 ppm of erucamide or oleamide.

LLDPE will normally require more slip additive than LDPE. LLDPEs produced with Z-N catalysts contain a higher concentration of low molecular weight amorphous species (waxes) which will absorb some of the fatty acid amide. The metallocene PEs have low extractables content and are easier to formulate.

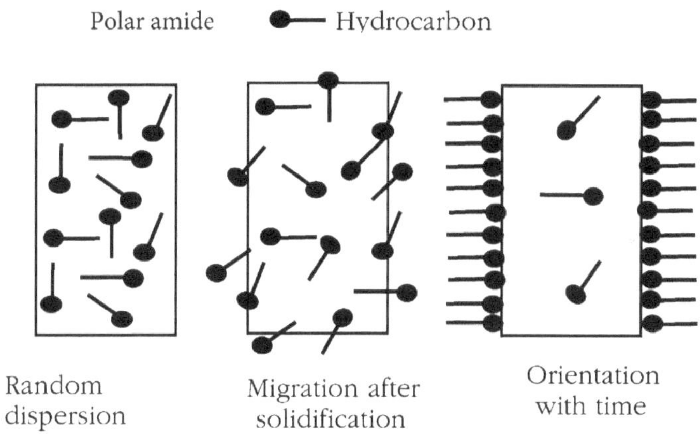

Figure 14.1

Slip migration mechanism

Erucamide and oleamide are the most widely used slip additives with polyethylenes. Oleamide has the fastest migration rate and is used with thin films running at high line speeds. Erucamide with its slower migration rate is preferred for specialty applications such as lamination films. Its slower migration rate enhances the corona treatment effect and reduces problems of migration into the adhesive after lamination.

Film thickness will have an important effect on the COF for any given concentration of slip agent. **Figure 14.2** shows the COF diminishing as thickness increases. At 100 μm thickness, the curves converge. Once the film's surface is saturated with erucamide any additional amount is wasted and can cause difficulties in printing or heat-sealing.

This thickness effect must be considered in coextrusion. Many processors will use slip formulated PE grades in all layers. For instance, if in a 2-layer structure each

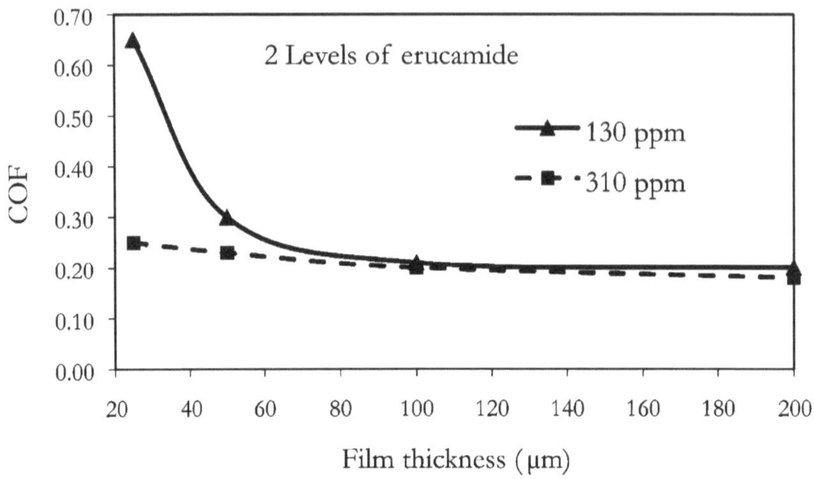

Figure 14.2

Effect of film thickness on COF

layer contains 500 ppm of oleamide the combined layers will contain 1000 parts, all of which will exude to the outer surfaces of the final film. This may prove excessive and affect sealability. In this example the slip should be reduced to 250 ppm per layer. Another option is to place the additive in one layer only and allow a few hours for the slip to migrate and equilibrate at the surfaces.

It is not recommended to add slip agent from a masterbatch. The PE is best supplied ready formulated by the manufacturer. Masterbatches can vary in quality and may contain dispersing aids and other impurities which affect adhesion and COF in lamination. There is also the possibility of mixing errors by the processor.

14.2 Antiblocking additives

Blocking is defined as the adhesion between adjoining layers of film. It may arise during processing or storage. Blocking is the force required to separate two layers of film after conditioning at a controlled temperature and pressure. With blown PE film it is important to be able to easily separate the layflat tube after extrusion. Bags and layflat tubing made from blocked films will be rejected by the converter and end user.

The antiblock additives must form hard projections on the film surface thus roughening it in a manner analogous to sandpaper. The amount will usually range from 500-2000 ppm. Higher levels of antiblock are required with LLDPE compared to LDPE. Copolymers such as EVAs also require high levels of antiblock agent. The following inorganic additives are used as antiblock additives in polyethylene films. At high comonomer content stearates (sodium, zinc) can be used.

- **Synthetic Silica (SiO_2)**: at particle size >2 μm synthetic silicas are efficient and will minimize film haze. Cost is high and these materials are very abrasive and may cause scratches.
- **Natural Silica**: these are cheaper than synthetic silicas and have a less favourable particle size distribution. Haze is higher than with synthetic silicas.
- **Talc**: tending to replace silica in some applications. Narrow die gaps should be avoided when using talc as an antiblock. Narrow die gaps will excessively orient the particles in one direction.
- **Calcium Carbonate ($CaCO_3$)**: highly efficient but only used when film clarity is not required.

When combined with slip agents all these antiblock additives have a very strong influence on COF. The efficiency of oleamide for instance is improved in combination with natural silica.

14.3 Some guidelines on slip and antiblock selection

- The faster migration of oleamide can be an advantage with in-line bag-making machines where a low COF is needed quickly. Rapid migration may also help to alleviate wrinkle problems caused by excessive friction against the collapsing boards of the blown film line.
- For the same ultimate COF, oleamide will generally require higher concentrations due to a higher volatilization (loss) rate as the bubble cools.
- Under adverse sealing conditions erucamide may give more unwanted plate-out on the sealing bars because of its lower volatility.
- Erucamide is preferred for adhesive lamination films because of its lower volatility.
- The slower exudation rate of erucamide makes in-line corona treatment easier.
- The slower exudation of erucamide may be of benefit in critical winding operations where too slippery a surface will cause problems such as roll telescoping.

14.4 Antistatic agents

Antistats work in polyethylene films by migrating to the surface. Once established on the surface, the antistat increases the surface energy of the polymer while decreasing the surface tension of a liquid, i.e. moisture, resulting in good wetting properties. The presence of moisture causes faster charge decay.

Ethoxylated amines, e.g. nonylphenol ethoxylate (NPE), are the most widely used antistatic additives in polyolefins. To be effective these amines must react with atmospheric moisture. Measured surface resistivity or decay time shows that effective antistatic response will occur only in the presence of at least 40% RH. A resistivity of 10^9 ohm is obtained with ethoxylated amines at the right conditions. If lower values are specified then carbon black will be required to further lower resistivity.

Slip, antiblock and antistatic additives are interactive and their simultaneous use can be unpredictable. Other additives such as light stabilizers, antioxidants, processing aids and pigments can also interfere with the efficiency of the antistat. Extrusion conditions and corona treatment may also influence the response.

An effective antistat must meet the following criteria:

- Fast charge decay
- Usable in transparent and colourless applications
- No effect on printability
- Active at low humidity levels

Other techniques such as using antistatic bars (ionization), carbon brushes, tinsel etc. placed before the winders are also used by many converters. Some use these simple techniques in conjunction with additives.

14.5 Antifog agents

Antifog agents are blended in polyolefin films to prevent or reduce the condensation of water in the form of small droplets which resemble fog. These agents include alkyphenol ethoxylates, complex polyol mono-esters, polyoxyethylene esters of oleic acid and sorbitan esters of fatty acids. They function as wetting agents, which exude to the film surface and lower the surface tension of water causing it to spread into a continuous film rather than forming droplets. The contact angle is reduced and the water will spread out on the surface of the film. These non-ionic surfactants are added as concentrates to the polyethylenes. Concentrations will vary from 0.5 to 5% addition depending on the film thickness and application. These antifog products work well in LL/LDPE and EVAs but less so with polypropylene. In coextruded films the antifog agent is added in the inner layer.

For food packaging such as for fruit and vegetables, concentrations of around 1-2% are recommended. In the agriculture sector antifog is required to prevent the formation of droplets on the internal film surfaces of greenhouse covers. The droplets will reflect light slowing down plant growth. Concentrations of 1 to 4% antifog are used depending on the film properties and climatic conditions.

14.6 Peelable seal films

The need for consumer friendly packaging has spurred the development of easy to open packs. Peelable seals were developed to meet this need in flexible packaging. Peelable seals are a contradiction for the packaging engineer since the critical function of the package is to protect the integrity and prevent abuse of the packed product. However, the consumer requirement for reliable easy to open packages needs to be met by the innovative film fabricator. A number of approaches have been developed over the years to meet the conflicting requirement of easy openability with product protection. Two basic techniques are used:

1. Lacquer coatings, which can be separated relatively easily or broken down at the seal.
2. Seal layers based on incompatible polymer blends, which can be coextruded, laminated or extrusion coated to the support.

Lacquers are traditionally used as a peelable sealant for lidding stock based on OPET, BOPP or aluminium foil. Since the coatings are very thin, irregular distribution of the sealant will occur if the seal flange of the tray is not perfectly flat. This will lead to variable seal strengths, which may be overcome by higher sealing temperatures, but can give uncontrolled peel strength resulting in tear or break of the lidding stock support. Consequently other options have been developed over the years.

Easy peel films are produced by modifying the resin's morphology with a material that inhibits the complete melting and fusion of its surface. This is usually accomplished by the use of incompatible blends. The intent is to create a seal of limited strength. A binary blend is produced by mixing the base polyolefin resin with an immiscible polymer. The immiscible material forms a finely dispersed phase in a continuous phase of the host polymer, typically LD/LLDPE. On heat-sealing the continuous phase will seal as normal. The dispersed polymer phase, however, forms a fracture prone interface, which will initiate separation or peeling when stressed. The basic morphology is shown in **Figure 14.3**. Obtaining the fine balance of seal integrity and easy peel is obviously critical in not only the formulation but also in the quality of mixing.

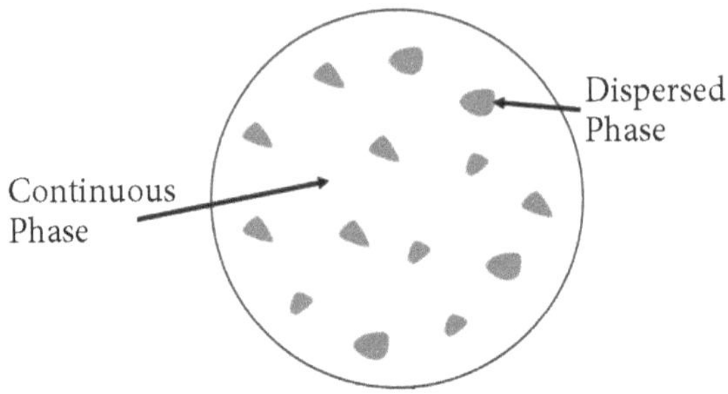

Figure 14.3

Morphology of 2-phase blend

Poly-n-butene (PB) is a useful polymer for such modification of LD/LLDPE polymers. Another suggestion is the use of SEBS (styrene-ethylene/butylene-styrene block copolymer) in similar blends.

Figure 14.4 shows the effect of adding varying amounts of PB to either EVA or LDPE. The zones of functional peelability are shown with either high PB or low PB concentrations.

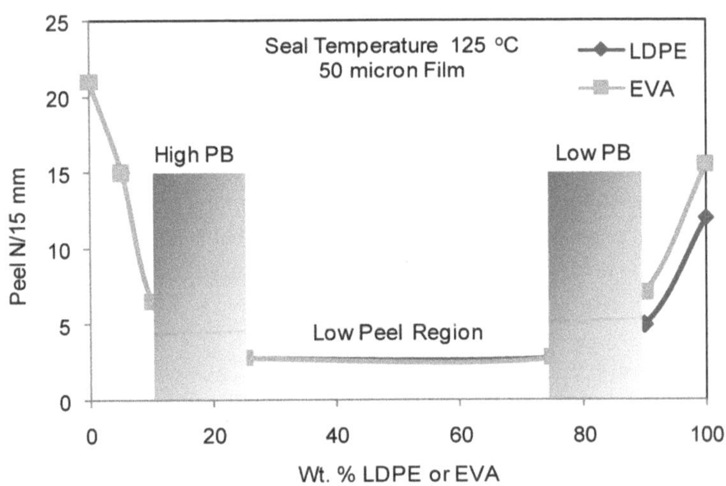

Figure 14.4

Peel/seal map for PB-modified EVA and LDPE

For blown LDPE or EVA films a 15% addition of PB is usually recommended. In cast films where the amorphous PE concentration is higher thus increasing the compatibility with PB, peel strength may be higher and the blend will need reformulating.

LLDPE has a higher level of compatibility with PB compared to the more branched LDPE. Furthermore, the higher melt viscosity of LLDPE will also improve the degree of mixing in the extruder. The overall effect is that the LLDPE/PB blend will develop stronger seals than the LDPE equivalent. This requires consideration when using either pure LLDPE or as more usually ternary blends with LDPE. The effect of adding LLDPE to a LDPE-PB blend is believed to improve the PB dispersion in the PE matrix. The size of the dispersed PB particles is reduced, i.e. refined, and results in reduced peel force. With pure LLDPE the result is reversed (higher peel forces) as the system is now more compatible in the absence of branched LDPE.

Compatibility with PB is further enhanced with metallocene PE. Therefore, higher levels of PB must be added to blends containing mPE. The indications are that to obtain a reasonable peel strength window with mPE, the amount of PB is increased to around 40%. The high cost of PB may make this an unattractive solution. Formulating PB with PE is clearly complex bearing in mind that other variables such as the melt index of the blend components will also play a role in the quality of the mixing, adding further complications.

Chapter 15

Corona treatment

Corona discharge treatment has for many years been an important weapon in the armoury of film converters. The technique is widely used to prepare surfaces for printing, laminating and coating.

Corona is generated by applying a high frequency, high voltage current to an electrode separated from an earthed surface by an air gap, a substrate and a layer of dielectric material. Frequencies are in the range of 10 to 50 kHz with peak voltages of 30 kV. When the voltage across the gap reaches 3,000 to 5,000 volts per millimeter (mm) the free electrons accelerate towards the positive electrode with energy high enough to displace electrons from molecules in the air gap. The outcome is a cascade effect with electrons and ions being produced, which results in a flow of current across the gap. Any variance that occurs is due to surrounding conditions such as humidity. Large ionization currents are produced and as the power input is increased the potential across the air gap remains more or less constant. This means that increasing power across the corona will proportionately accelerate the rate at which the charged particles flow. For efficient treatment switch mode power systems are used. These provide a quantitative measure of the match between the load and the corona power. These switch mode systems makes it possible to install display panels that show the operators true output conditions.

When a dielectric material such as a plastic film is placed in the path of these particles, surface changes will take place on the plastic. The accelerated electrons on the surface of the film will cause the polymeric molecules to rupture and form reactive sites. The ozone generated by the discharge will oxidize these sites thus increasing surface tension.

The electrode or applicator consists of one or more metal electrodes spaced round the circumference of a metal roller covered with a dielectric. The dielectric sleeve must be made from an ozone resistant material and completely free of cavities or micro-holes, which may cause breakdown during operation. The material must also have a low dielectric loss to minimize heating. Roll covers can be made from vulcanized elastomers or ceramic materials. The ceramics are less prone to decomposition from the discharge energy than organic elastomers.

The optimum air gap depending on the dielectric material and the frequency is in the range of 1-2.5 mm. Electrode design is critical. Treating efficiency is improved by the use of multi-electrode configurations to spread the corona discharge over the widest area.

The oxidized species formed on the polyethylene surface increases its surface free energy rendering it more receptive to wetting by liquids. It is important to remember that the plasma penetrates the surface to a depth that is unlikely to be greater than a micron. This means that the treatment can be rubbed off by friction, since polyethylene has low abrasion resistance. Loss of treatment, in certain cases, can be attributed to abrasion from faulty idler rolls on the converting machine. Softer films such as EVA copolymers and POEs are more prone to this rubbing-off effect.

Elemental analysis of PE film surfaces by ESCA techniques confirms the presence of oxidized species following corona treatment. The data in **Table 15.1** was obtained on an LLDPE polymer free of slip/antiblock additives.

The small amount of oxygen found on the untreated sample is probably due to the presence of impurities in the polyethylene. Treatment clearly increases oxygen concentration, which tends to drop slightly with ageing. The film surface wettability did not, however, change and the film could still be readily printed.

Table 15.1
Loss of oxidized species with time

	Day 1		12 Months	
	C	O_2	C	O_2
Untreated	99.2	0.8	99.4	0.4
Treated	86.9	12.2	91.0	8.6

In-line corona treatment is preferred for two reasons. The first is that the more crystalline the polyethylene film the more difficult it is to treat. Since polyethylene films will increase in crystallinity on ageing, they will require more generator power after a day or two for a required treat level. This effect is minimal with films produced by the blown film process that do not undergo significant crystallinity changes (slow cooling) with time. It may however be significant with cast films that are processed at higher temperature and rapidly quenched on a chill roll.

Secondly, the films will usually contain a slip additive that will migrate to form a layer on the surface shielding the polyethylene from the corona plasma. In **Figure 15.1**, an LDPE film containing slip additive was corona treated in-line and the surface tension measured over a series of time intervals.

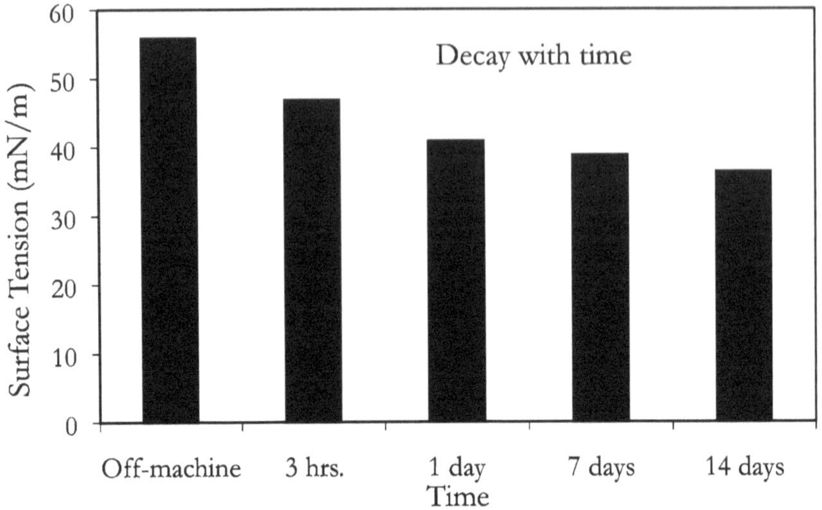

Figure 15.1

Effect of slip agent migration on treatment

The surface tension drops from 56 to 37 mN/m. The biggest drop occurs within the first 24 hours. In **Table 15.2** the samples from **Figure 15.1** were retained and the surfaces washed with isopropanol. The surface tensions were remeasured. The initial surface tension is regained after washing.

Table 15.2

Effect of solvent wash on treatment (Initial treat level= 56 mN/m)

Time	Unwashed (mN/m)	Washed (mN/m)
3 hours	47	56
1 day	41	56
7 days	39	56

In spite of these results, many converters complain of poor adhesion in printing or laminating of treated PE films usually after ageing. These complaints are more prevalent with film rolls that have been stored for some undetermined time and climatic conditions. When this happens there is no other option but to repeat the corona treatment.

Table 15.3 illustrates the difficulty of corona treating a film that has been aged over a period of time. In these experiments the films were post-treated one day and one week after extrusion. The lower treat levels achieved after the elapsed times, compared to the off-machine value, are caused mainly by the screening effect of the slip additive as it migrates to the surface. Increased crystallinity with time may also contribute to the lower treat levels. The drops are, however, not very significant after the first 24 hours.

Table 15.3

Post-treatment of film

Time	Treat level (mN/m)
Off-machine	56
1 day	50
7 days	49

The dependence of corona treatment effectiveness on slip agent (erucamide) concentration is plotted in **Figure 15.2**. The slip level was varied from zero to 1000 ppm in 50 μm thick film. The line speed was fixed at 28 mpm and the corona generator was maintained at the same power setting throughout these experiments.

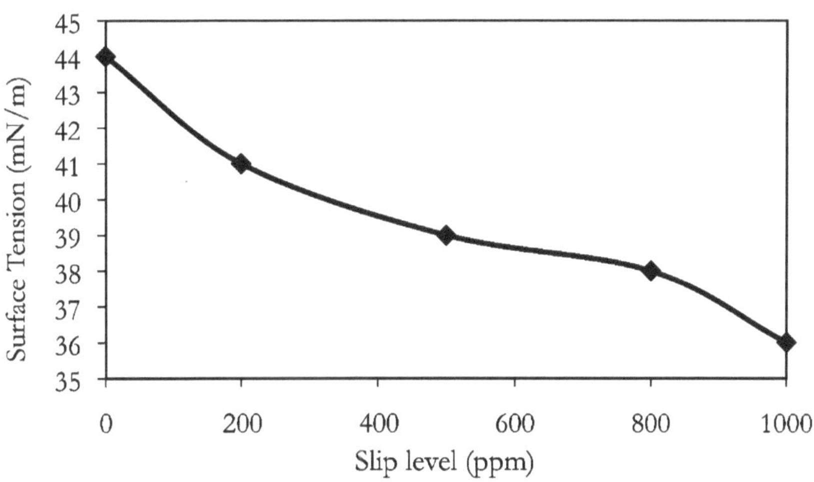

Figure 15.2

Effect of slip concentration on corona treatment

As can be seen from **Figure 15.2**, treatment becomes more difficult as the slip agent concentration increases.

As a conclusion it is important for the converter to select a corona treater of sufficient power for the throughput of the film line and ensure that the supplier provides the optimum design of electrode configuration. Additives, especially fatty acid amides and stearates will have a negative effect on treatability and these must be accurately controlled to avoid poor ink adhesion or inadequate bonding with lamination adhesives. Corona treatment should be applied before the additives have had time to migrate to the surface.

Table 15.4 shows some recommended corona treatment levels with different types of printing inks on polyethylene films.

Table 15.4

Corona treatment level for PE films

Printing process	Treatment (mN/m)
Flexographic/gravure	
Water	38-44
Solvent	36-40
UV	38-50
Lithographic	
Water	40-46
Solvent	37-42
UV	40-50
Offset/letterpress	
Water	40-46
Solvent	37-42
UV	42-54

15.1 Other treatment methods

Flame treatment can also be used to increase the surface tension of PE and PP films. The process is claimed to reduce the formation of pinholes in thin films and produce a more even treatment across the web. Key parameters to consider are:

- Air to gas ratio: propane or butane is used at a recommended ratio of 24-28:1
- Gap: 10-12 mm
- Gas flow: ~ 700 l/min
- Line speed: effectiveness will drop with increasing speed, as with corona treatment
- The applicator may have ribbon or drilled port burners

More recently new techniques have been developed to improve the efficiency of film laminating processes. High density atmospheric plasma represents a new generation of surface treatment that is claimed to be ideal for film lamination. Atmospheric plasma involves the electrical ionization of a gas, similar to corona treatment. The plasma creates a cloud of ionized gas that reacts with the substrate surface to raise its surface energy and improve adhesion. The process operates at lower voltage levels than corona discharge. An AC power source is used with a proprietary electrode, a dielectric between the electrodes and a selected gas mixture as the plasma medium. Helium and acetylene are suggested gas mixtures for plastic film applications. Other gases, such as oxygen (O_2), nitrogen (N_2) and carbon dioxide (CO_2) can also be used.

With nitrogen, nitrile functional groups are formed, which can form strong bonding sites. The absence of oxygen prevents oxidation and chain cleavage. An interesting claim is that the equipment can be adapted to deposit nanocoatings of organosilicates and organometallic compounds unto moving webs. These nanocoatings are claimed to improve barrier and adhesion properties and the following advantages are claimed for atmospheric plasma vs. corona treatment:

- More durable treatment
- Superior surface energy and uniformity (treatment >60 mN/m possible)
- No backside treatment with low gauge films
- Porous materials can be treated with no pinholes
- Low level of ozone generation
- Treatment without oxygen
- Does not need a treater roll
- Lower voltage

The following advantages are claimed for plasma versus flame treatment:

- Can be used to treat heat-sensitive webs
- Chemistry of surface can be tailored per application specifications
- Less fire hazard
- No overtreatment

The cost of the reactive gases and higher equipment capital investment could be a limiting factor in the wide use of this technique.

Table 15.5 compares the changes in surface tension of polypropylene films for three treatment methods. The lower the contact angle the higher the surface tension (best wettability). The atmospheric plasma is the most effective from this data. Corona appears to be the least effective.

Table 15.5

Contact angle on PP film

Treatment	Contact angle
None	67°
Corona	58°
Atmos. plasma	50°
Flame	57°

Chapter 16

Form, fill and seal packaging

One of the most versatile methods of packaging solids, powders and liquids with flexible films is with the use of form, fill and seal (FFS) techniques. These processes are fast, economical and hygienic for edible and pharmaceutical products. Many of the films used in FFS packaging are based on multilayered structures produced by coextrusion, lamination and extrusion coating.

FFS processes produce flexible packages from roll fed stock. PE resins play an important role since they contribute to the critical function of forming strong heat-seals on which the package's integrity will depend. Supermarket shelves are inundated with examples of goods packaged on FFS machines.

There are two basic types of form, fill seal processes:

- Vertical form, fill and seal (VFFS)
- Horizontal form, fill and seal (HFFS)

Figure 16.1 shows the basic layouts of the HFFS machines. There are two set-ups; the first uses a single web, which is folded as it moves downstream and filled. The formed pouch is 3-side sealed. The second method uses two webs, which cover the product and are 4-side sealed. This technique works very well with thermoformed trays, which are produced in-line, and are easily filled and then lidded with the second web.

VFFS processes are excellent for the production of simple pillow style pouches or gusseted bags. The latter create more volume and maintain a rectangular shape after filling. The web is folded to form a tube and sealed vertically where the edges meets. The vertical seal can be a fin or lap seal and the pack may be gusseted as shown in **Figure 16.2**. The type of seal used places different demands on the properties of the film. A gusset will require the heat from the sealing bar to pass through four layers of film, which will slow the process unless a very low seal initiation temperature (SIT) material is used. In the case of the lap seal it is critical that both the inside and outside of the film are sealable to each other. Although not a problem with monofilms this may create difficulties with asymmetric multilayer structures.

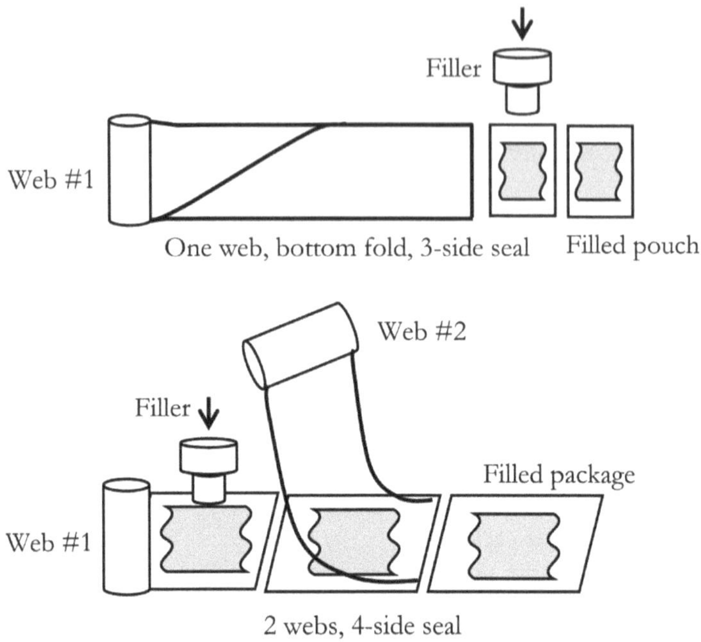

Figure 16.1

HFFS Packaging layout

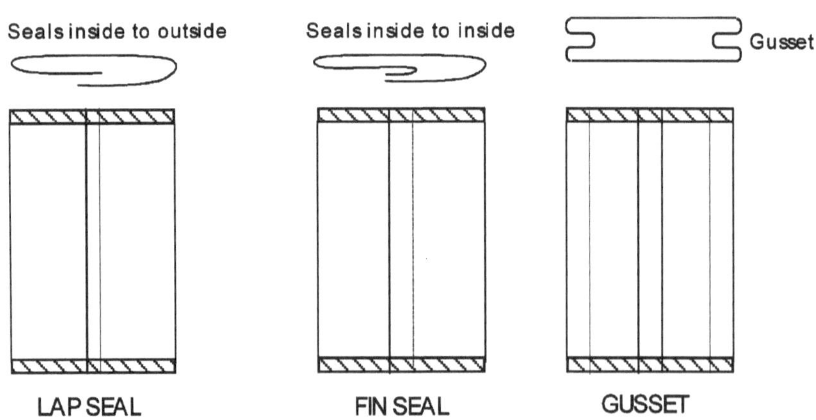

Figure 16.2

Seal types in VFFS systems

Figure 16.3 is a schematic of the basic VFFS layout using a single web. The operational steps are as follows as the tube is folded and sealed vertically to form the tube parts:

1. Filling tube.
2. Forming collar to fold web.
3. Tracking roller, which feeds the web.
4. Vertical sealing bar, which forms the tube.
5. Web transport belts.
6. Horizontal seal bar.
7. Cutting knife.

In the final step (7) the seal bars at the bottom are removed and the pouch or carton is cut and separated. There are variations on this basic process which use two webs to form the tube section. These packs are 4-side sealed.

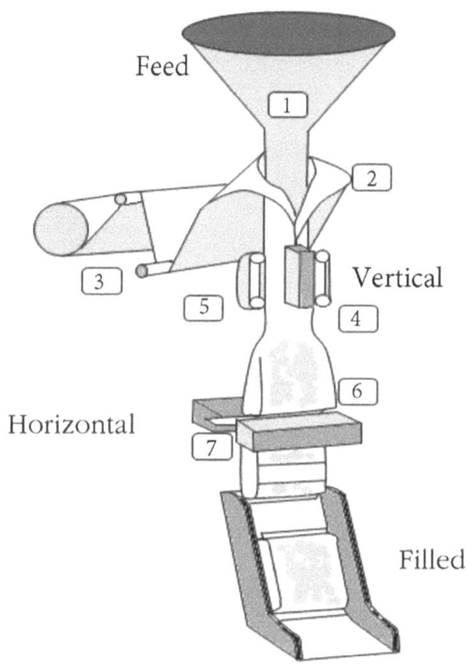

Figure 16.3

VFFS Packaging machine with single web

To avoid opening the bottom of the pouch as it fills under gravity sufficient time must be allowed for the hot seal to cool. It is therefore important for the melt to have high enough hot tack to resist opening from the applied stress during filling. Laminates that include an aluminium layer will have a much faster cooling rate enabling higher throughputs.

High/radio frequency (HF/RF) sealing can be used with foil bearing structures particularly with thick paperboard supports. The conductive metallic layer is heated rapidly by induction and melts the sealant layer. This technique bypasses heat dissipation through the thick paper and accelerates the process. Cartons used in liquid packaging will normally use induction sealing.

The VFFS process is low cost and is very versatile in terms of pack sizes and product changes. Production speeds of 300 pouches/min. are achieved with some machines. However the process is more restricted in the choice of width to length ratio of the pouch compared with HFFS packaging.

All the PE polymerization processes based on organometallic catalysts will produce polyethylenes with narrow molecular weight distributions. These will convert into films with superior seal strength and hot tack compared to branched LDPE and EVA. Lowering the density < 0.915 g/cm^3 with metallocene catalysts will lower seal initiation temperature proportionately and improve hot tack.

For the converter it is important to select the heat-seal layer on the basis of cost, faster packaging rates, reduced leakers and reductions in other packaging failures.

In applications where sealing through contaminants such as bacon fat and fine powders is most challenging the ionomers have generally proven to be the material of choice. **Figure 16.4** shows the comparative effectiveness of an ionomer

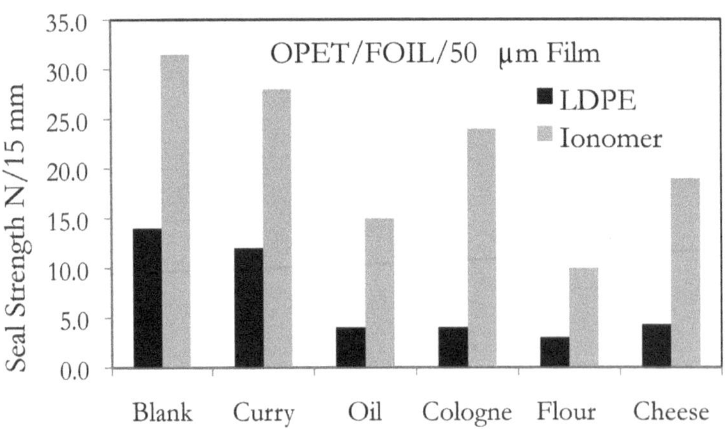

Figure 16.4

Effect of heat-seal layer on seal through contaminants

in heat-sealing through a range of consumer products. The LDPE and ionomer were applied on an OPET/aluminium laminate. The superiority of the ionomer (Zn^{++}) is clearly shown. The seal layers were 50 μm thick.

Figure 16.5 shows similar data on the comparative sealability of a number of copolymers through bacon fat and milk powder. The best results are with the EAA and the ionomer.

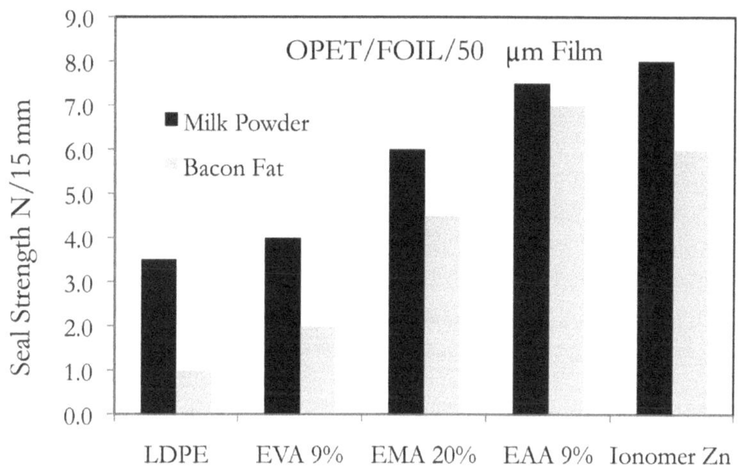

Figure 16.5

Seal strength of various copolymers through milk powder and bacon fat contaminants

The following critical polyolefin film properties are applicable to form, fill and seal processes for both liquid and dry goods packaging:

- Hot tack strength
- Seal strength
- Low seal initiation temperature (SIT)
- No pinholes
- Low COF between film and metal
- Tear and puncture resistance
- No taint/odour (food packaging)
- No light transmission for dairy products
- Softness and flexibility (pillow packs)

- Peelable seal (easy opening)
- Seal through contamination (fat, powders etc.)
- No static (dry goods, fine powders)
- Low moisture vapour transmission rate (MVTR) for cereals

For the best results on all FFS machines, a narrow film thickness distribution is required, typically ±8% from point to point. The average value should equal the target thickness.

16.1 Calculating film stiffness

The handle of a film on packaging and filling machines is difficult to quantify. The film should ideally be non-extensible and have sufficient rigidity to be pulled and wrapped without wrinkling and creasing. With the current emphasis on downgauging the risk of producing films that will stretch or be too flexible for machine handling has increased. The calculation of stiffness can be a useful measure of assessing the ability of a film to run on FFS converting processes.

The stiffness or flexural strength S is defined by the product of the elastic modulus E and the moment of inertia (J).

$$S = E \times J \tag{16.1}$$

The moment of inertia (J) is calculated from equation 16.2 using the thickness and width of the film.

$$J = \frac{w \times t^3}{12} \tag{16.2}$$

w = width t = thickness
Film stiffness (S) is calculated as:

$$S = E\frac{w \times t^3}{12} \tag{16.3}$$

Equation 16.3 shows that the stiffness (S) is proportional to the elastic modulus (E) and the cube of the specimen thickness (t). This relationship clearly demonstrates the importance of thickness in determining the overall stiffness of films. Typical moduli and stiffness values of some PE films at 30 μm thickness and 1 m width are shown in **Table 16.1**.

Table 16.1
Stiffness of PE films (30 μm)

Resin	E-Modulus (MPa)	Stiffness (S) ($N.mm^2$)
mPE (0.905 g/cm^3)	110	0.25
EVA (5% VA)	130	0.29
Ionomer (Zn^{++})	160	0.34
LDPE (0.919 g/cm^3)	160	0.36
LLDPE (0.918 g/cm^3)	200	0.45
Ionomer (Na^+)	280	0.63
MDPE (0.940 g/cm^3)	550	1.24
HDPE (0.952 g/cm^3)	800	1.80

In **Figure 16.6** the exponential increase in stiffness with thickness is shown for a 1 m wide web. Thickness clearly will have a marked effect on the machineability of the film. In **Figure 16.7** several ethylene polymers are compared for stiffness versus film thickness. The stiffness of the POE type mPE is close to that of a low EVA (5%VA) copolymer. The ionomer is stiffer than LLDPE. HDPE is almost an order of magnitude higher than the polymers at the low end of the scale in this table and hence HDPE is a useful material to increase stiffness by either blending or coextrusion with LL/LDPE. For maximum stiffness HDPEs with densities >0.950 g/cm^3 should be used.

In coextrusions placing the higher modulus materials on the outside will maximize the overall stiffness of the structure.

The thinner the film the faster is the sealing and cooling rate. It is also important to bear in mind that the elastic modulus will depend on crystallinity and the method of processing. For instance cast films will be less rigid than blown films. Orientation can also increase the elastic modulus by around a factor of 2.

Figure 16.8 compares the stiffness of a number of films of typical commercial thicknesses for flexible packaging. BOPP at 20 μm has the highest value followed by HDPE. The 12 μm OPET is close to the BOPA (oriented polyamide) at 15 μm. The 20 μm PA6 is a cast film. These calculations all emphasise the significant effect that the thickness and polymer selection has on the overall stiffness and handle of the film.

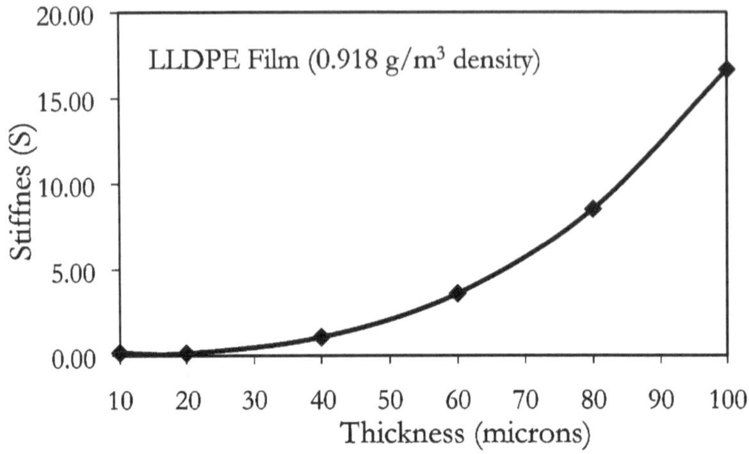

Figure 16.6

Stiffness vs. thickness for LLDPE film

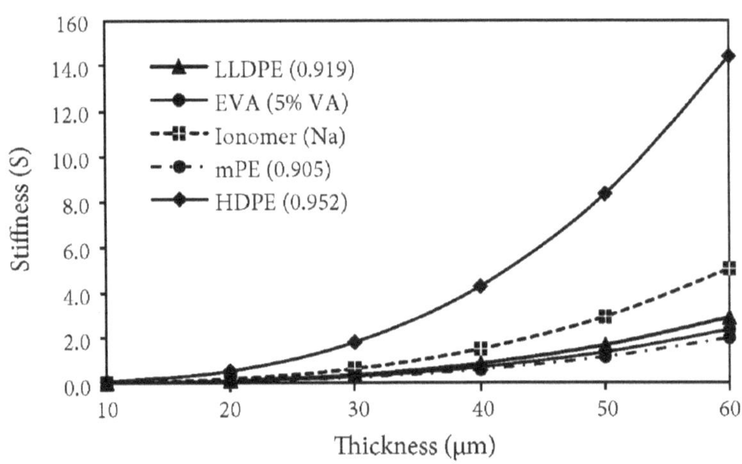

Figure 16.7

Stiffness vs. thickness for various ethylene polymers

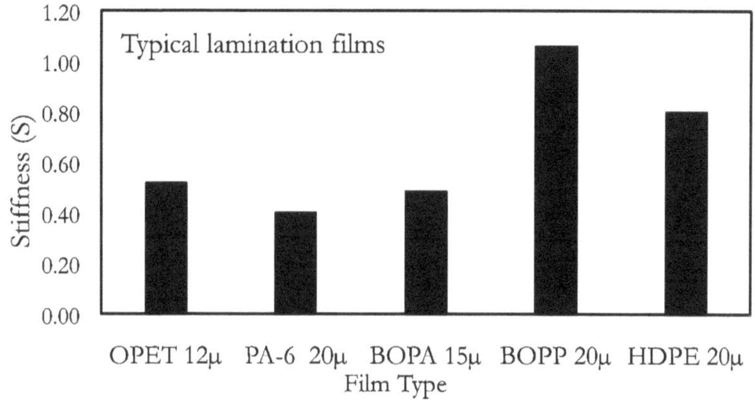

Figure 16.8

Stiffness of various packaging films at commercial thicknesses

Chapter 17

General purpose polyethylene films

This grouping of polyethylene film applications covers a very wide range of household and industrial uses. The market is cost driven. Resin selection will very often be based on the use of blends to balance processability, properties and cost. The low cost LLDPEs (butene) are very often blended with both LDPE and HDPE resins. Some examples of applications are listed below:

- Grocery and shopping bags
- Bin liners
- Household garbage bags
- Laundry bags
- Textile bags
- Sandwich bags
- Freezer bags
- Liners
- Packaging of paper tissue products
- Mailing envelopes
- and others

17.1 Garbage bags

This grouping includes garbage bags used in households and in industry. The bags are marketed in packs of gusseted film on the roll with an easy tear serrated edge. Typically, bin liners once filled are placed into the larger (30 L) household/industrial garbage bags and disposed of. The garbage bags then end up in the waste bin for institutional collection. **Table 17.1** compares LD/LLDPE with high molecular weight (HMW) HDPE bags.

Table 17.1

Garbage bags from LD/LLDPE and HMW-HDPE

Polymer	Thickness (μm)	Dart (g)	Dart (g/μm)
LD/LLDPE	45-70	80-120	~1.7
HMW-HDPE	20-30	150-200	~7.0

Typical properties of HMW-HDPE municipal refuse bags found in Europe are reported in **Table 17. 2.** HDPE is rapidly replacing LDPE.

Table 17.2

Average properties of HDPE garbage bags

Properties	Unit	Type #1	Type #2
Average thickness	μm	28	22
Tolerance	±%	12	18
Tensile strength MD/TD	N/mm^2	48/40	45/39
Elongation MD/TD	%	420/660	320/630
Dart drop	g	120	70
Dart drop	g/μm	4.3	3.2
Bottom weld (% of tensile)	%	70	70
Side weld (% of tensile)	%	65	70

Coextruded bags are sometimes used in the institutional bag segment, where LL/LDPE outer and inner layers give the required puncture resistance using a centre layer of scrap and recycled material. For both cost and environmental reasons the garbage bag market has been an active player in the use of recycled materials. The challenge is to source recycled material of consistent quality. Typical 20-30 litre refuse bags will contain 50% recyclates. Film thickness will range from 20-35 μm. The recyclates will usually be blended with LLDPE.

For the lowest carbon footprint, bags made from 12 μm virgin HMW-HDPE films are the preferred option[1]. This study includes comparisons of bags processed from LLDPE and biodegradable materials. The latter are costly and surprisingly have a higher carbon footprint than 12 μm HDPE. Resource reduction by downgauging seems to be the best approach in minimizing carbon footprint.

Typical sources of recyclates include the following:

---

[1] Detzel, A., Wellenreuther, F., Kunze, S., LCA of waste bags, **2009**, Institut für Energie- und Umweltforschung (IFEU) Heidelberg (http://www.plasteurope.com/news/WASTE_BAGS_t214249)

Source	Type
Agriculture films	Black, white or green
Shoppers/carrier bags	White and other colours
In-house scrap	Mixed colours

When using recyclates from such varied sources it is important to use black masterbatches that can effectively mask the variety of colours and also help in heat-stabilizing the total blend. Masterbatch selection should be based on both its covering power and colour strength. Usually 2-2.5% carbon black is required in the final let-down, however for best results the type of black, quality of dispersion, stabilization and diluent used will play a large part in obtaining the desired masking effect and acceptable physical properties. Costs can, therefore, be minimized by careful selection of the highest quality masterbatch and not necessarily buying the cheapest.

Other sources of recyclates are post-consumer PE films from shrink wrap, dry cleaning film, merchandise bags, grocery sacks, stretch wrap, recycle bags, bubble packaging film, commercial/industrial liners/bags and commercial overwrap.

17.2 Bin liners

This market sub-segment is defined as very low thickness bags placed into bins for waste collection. The main purpose of the liner is to protect the bin from dirt and refuse. **Table 17.3** compares low gauge HMW-HDPE bags used in non-institutional refuse applications, such as household and office trash can liners. HDPE has virtually completely replaced LDPE in this segment. Further downgauging to 9 µm is claimed.

Table 17.3

Properties of HMW-HDPE liners

Property	Unit	Typical values	
Average thickness	µm	10	10
Tolerance	±%	14	19
Tensile strength MD/TD	N/mm^2	62/60	62/46
Elongation MD/TD	%	320/530	290/490
Dart drop	g	80	48
Dart drop	g/µm	8.5	4.8
Bottom weld (% of tensile)	%	80	80
Side weld (% of tensile)	%	70	70

17.3 Shopping and carrier bags

The polyethylene shopping and carrier bag was first developed in Sweden. The initial design was based on the idea of forming a one piece bag made from a flat tube of PE film by folding, heat-sealing and die cutting to form the bag. The original patents date from 1965. These bags rapidly eroded the use of paper bags particularly in Europe where paper was relatively costly compared to LDPE. The reverse situation existed in the US.

The development of bags with easy-to-carry handles, high quality printing and graphics and low weight established their pre-eminence in this market. Resource reduction by downgauging with HMW-HDPE films has been an additional driving force. These water resistant bags also have the advantage of being reused as bin liners for trash collection.

This segment covers a variety of uses from simple T-shirt type grocery bags to more sophisticated designs of bags for use in high fashion retail outlets. In the USA, this market segment is classified as merchandise bags. The following design descriptors are used:

- T-shirt bags
- Patch handle bags (die-cut handles with patch reinforcement)
- Die-cut handle bags
- Snap handle bags (rigid moulded handles)
- Drawstring bags
- Sine wave bags (pull bags)
- Flat bags with no handles
- Bags with injection moulded handles

Originally, the market was exclusively blown LL/LDPE film, which has gradually been eroded by HMW-HDPE bags. Bimodal HMW-HDPE made it possible to downgauge from 55 to 22 μm film. Further reduction to 12-14 μm is now possible with the long neck blown film process. HMW-HDPE is widely used for grocery bags of the T-shirt type at supermarket checkout counters in Europe, the USA, Japan and other regions. Some jurisdictions are banning retailers from giving away free shopping bags at check-out points.

T-shirt bags from HMW-HDPE are functional and cheap and tend to be favoured in supermarkets and discount food stores. In other situations where brand or store image is not important the low cost T-shirt HDPE bag is again the preferred choice. Bimodal MWD HDPE grades offer the best opportunity for downgauging.

Resource reduction is the best way to reduce carbon footprint. Bimodal grades will also more readily accept recyclates which can further reduce overall raw material cost.

Three part blends of 60% LDPE, 25% LLDPE and 15% MMW-HDPE produce excellent quality films for applications requiring high strength at moderate stiffness. These blends have the advantage of being extruded at high outputs on LDPE blown film lines. The addition of colour masterbatch will increase the raw material cost by an average of 8%.

This market is cost driven and highly competitive. Some processors have developed coextruded formulations as a means of driving down costs. These consist of blends of HMW-HDPE with LL/LDPE. These are added to improve heat-sealability and reduce cost with LLDPE. The LL/LDPE content is usually between 15-30%. Calcium carbonate (~5%) is also added to improve openability and increase opacity. The inorganic filler is added from a masterbatch.

Some data comparing the properties of different PE resins used in T-shirt grocery bags is shown in **Table 17.4**. Measurements were carried out on actual commercial bags available in supermarkets.

Table 17.4

Properties of T-shirt bags

Property	Units	LD/LLDPE	LDPE	HMW-HDPE
Thickness	μm	26	34	20
Dart drop	g	38	84	146
Dart drop	g/μm	1.46	2.47	7.3
Tensile yield MD/TD	MPa	11.6/12.6	9.7/10.7	33.3/29.3
Tensile @ break MD/TD	MPa	30.7/17.2	17.6/16.2	39.0/37.6
Elong. @ break MD/TD	%	120/610	430/619	225/425
Elmendorf tear MD/TD	g/μm	11.4/10.4	4.8/23.8	1.0/7.2

Table 17.5 shows data comparing a series of polyethylenes recommended for carrier bag applications.

17.3.1 Reusable shopping bags

In many jurisdictions laws have been passed banning single use plastic shopping bags and replacing them with reusable bags as a drive to reduce litter and carbon footprint. These are made from a range of materials that include canvas, jute, reinforced paper and synthetic fibres, usually polypropylene, as well as stronger, thicker plastic films. Although a good idea an interesting polemic has developed.

The reusable bag requires more energy to produce than the low gauge HDPE bag. One typical reusable bag apparently utilizes the equivalent amount of energy to 28 polyethylene shopping bags. Of course it is difficult to estimate how many reusable bags will be purchased annually by the average household and how this will convert into 12 μm HDPE bags. Health issues have also been raised on the basis that the reusable bag is not washed, which may lead to food poisoning. This is attributed to repeated exposure to raw meat and produce with an increased risk of bacteria and mould formation. Food waste is an issue in the advanced economies.

Another observation is that the single use T-shirt bag has a secondary use in collecting kitchen waste, knotting the filled bag and placing it the dustbin for collection. This avoids any cross contamination and the need to wash the kitchen bin. The alternative is to purchase PE bags on the roll at extra cost for the consumer which seems to be self-defeating. The debate is ongoing.

Table 17.5

Comparison of PEs for carrier bags

Property	Units	LDPE	LD/LL (C_4) 75/25%	MDPE C_6	MDPE C_6	HMW HDPE
Density	g/cm^3	0.922	0.922	0.930	0.937	0.955
Yield stress	MD/ TD MPa	12/11	11/12	17/16	21/21	35/30
Tensile strength MD/TD	MPa	20/28	25/20	50/35	49/47	58/52
Elongation MD/TD	%	200/400	200/500	300/550	400/600	303/500
Dart drop	g	150	120	175	182	160
Dart drop	g/μm	3.8	3.4	5.0	10.1	10.7
Thickness	μm	40	35	35	18	15
Tear strength MD/ TD	N	30/32	30/32	35/35	28/28	24/20
Orientation MD/TD	%	90/10	96/4	87/13	66/34	70/30
Extruder type (PE type)		LDPE	LDPE	LDPE	HDPE	HDPE
Extrusion BUR		2	2	3	3.5	3.5
Freeze line		Low	Low	Medium	High	High

17.4 Consumer and household bags

The household films and bags are consumer bag and wrap products that are purchased in retail outlets such as supermarkets. Household bags include a variety of storage bags, sandwich bags, kitchen bags and freezer bags. Whilst these are distinct categories of product some cross-usage does occur. The use of reclosable bags in this sector is gaining market share.

The household stretch wrap market is more mature and has been historically dominated by plasticized PVC and Saran (PVDC) wrap. The use of PE-based cling films has been encouraged as a replacement for vinyl. These initially tended to be EVA-based which aimed to replicate the extensibility and transparency of vinyl films. Cast LLDPE films are also used in this market. Other forms of kitchen wrap such as aluminium foil are also still used.

Reclosable (zipper) bags first introduced by Minigrip more than 50 years ago can be produced by the following two techniques:

1. The film and zipper are extruded in a one piece construction from a tubular film.
2. Film and zipper are extruded separately and combined by heat-sealing or other technology.

The category of bag termed breakfast bags in Germany and sandwich bags elsewhere have become standard kitchen commodities. The bags are supplied on ready to tear rolls. **Table 17.6** compares the properties of such bags found in the some European countries. The films can vary from very low gauge HDPE films to 15-20 μm LDPE films. The data in **Table 17.6** is based on specimens from different European converters.

Table 17.6

Average properties of kitchen bags

Property	Unit	Typical European specs.	
		HDPE	LDPE*
Average thickness	μm	9-10	18
Tolerance	±%	15	10
Tensile strength MD/TD	N/mm^2	65/55	38.7/25.1
Elongation MD/TD	%	330/550	90/750
Dart drop	g	54	<28
Dart drop	g/μm	6	<1.3
Bottom weld (% of tensile)	%	68	58
Side weld (% of tensile)	%	75	72

** may include LL/LDPE blends*

17.5 Freezer bags

Household freezer bags are used in the kitchen for deep freeze storage of foodstuffs. These bags come in various sizes and are supplied on the roll and made from 30-35 µm films. The films are based mainly on LDPE/LLDPE blends or coextrusions with an HDPE core layer. EVA resins are being replaced on a cost basis by LLDPE blends. Many suppliers of deep freeze bags claim usage from deep freeze conditions (-20 °C) to boiling.

The determining factor in this segment is cost. Where mono-films are specified, these will be blends of 30-40% LLDPE in LDPE or low EVA. The LDPE is typically MI 0.6 dg/min. (D 0.920 g/cm^3). The lowest cost LLDPE (butene) is normally used. For applications intended to resist hot water, linear rich blends are preferred.

Freezer bag sizes found in European supermarkets are typically the following sizes:

- 16.5 × 25 cm for 1 litre capacity
- 25 × 32 cm for 3 litre capacity
- 30 × 45 cm for 6 litre capacity

Three layer coextruded films are penetrating the freezer bag market. The properties of typical examples of coextruded films found in Germany and the UK were evaluated and are shown in **Table 17.7**. These results of a private study are based on 100 specimens from different retail outlets believed to be representative in the two countries.

Table 17.7

Properties of coextruded freezer bags (3-layer films)

Property	Unit	Germany	UK
		PE*/HDPE/PE*	PE*/HDPE/PE*
Average thickness	µm	35	30
Tolerance	±%	6	8
Tensile strength MD/TD	N/mm^2	34/28	37/36
Elongation MD/TD	%	320/650	330/620
Dart drop	g	105	90
Dart drop	g/µm	3.1	3.0
Bottom weld (% of tensile)	%	55	73
Side weld (% of tensile)	%	67	59

**LDPE or LL/LDPE blends.*

These coextrusions are expected to be usable from deep freeze to boiling water conditions. LLDPE-based sealant layers are preferred to cover the widest heat resistance range.

17.6 Garment and textile films

Textile bags are used to package items of clothing normally displayed in retail outlets. They cover and protect the apparel during transport, storage and display. Predominantly, the PE bags are made from clear and glossy films and will usually carry a printed logo and/or safety instructions. Thickness range is between 20 and 100 μm.

The PE film grades have to meet the following criteria:

- High gloss
- Low haze
- High slip
- Low gels
- Printability

For textile packaging appearance is important and polyethylene is being replaced by cast PP (CPP) film, which offers greater sparkle and clarity. The development of copolymers and coextrusion options make cast PP a very attractive option.

17.7 Mailing envelopes

This market for mailing of magazines, journals and advertising material has grown very rapidly in recent years. The use of paper envelopes has virtually disappeared from this application. The product to be packed is usually flat and fed to a pre-formed bag, which is then sealed ready for delivery. Some packagers may pass the envelopes through a shrink tunnel to improve appearance. It is important that the item remains flat, therefore the degree of shrink and shrink force must be minimal.

LDPE films dominate this market. Butene LLDPE is blended to reduce cost. General purpose grades of 1-2 dg/min melt index and density of 0.920-0.923 g/cm^3 are used. The films should have excellent seal properties and tear relatively easily. To facilitate sealing and minimize shrink tension some converters use 3-4 dg/min. melt index resins. These will produce films with more MD orientation, which reduces tear strength in that direction. Film thickness is typically 25-30 μm.

Chapter 18

Food packaging

FFS (form, fill and seal) machines are used to package most types of food products. These can be in liquid, powder or granular form. Typical products include milk, beverages, condiments, cereals, spices, frozen produce and snacks amongst many others. The multilayer films reviewed in this section are based on a polyolefin layer laminated or coextruded with other materials such as OPET, BOPP, aluminium foil, polyamide (PA), metallized films and other barrier films. The PE-based films used for lamination are mostly coextrusions designed to minimize cost and optimize properties.

For goods requiring moisture and oil protection HDPE is used. HDPE with density >0.955 g/cm^3 is used in coextrusions with LD/LLDPE, EVA or ionomers depending on the required heat-seal specifications. The metallocene PEs are also used in this market as they provide outstanding heat-seal, low temperature toughness and low extractables.

The trend is to downgauge in this highly competitive market. Blending combined with coextrusion is usual to optimize film properties at minimum cost.

Tear and impact resistance is best with EVA, ionomer and mLLDPE. It is also acceptable in many applications with LD/LLDPE, but poor with HDPE unless biaxially oriented.

Taint and odour are influenced by the quality of the PE grade, by the extrusion conditions (temperature, oxidation) and by the level of corona treatment which can cause oxidation. LDPE which has no catalyst residues or stabilizers is usually the most organoleptically neutral of the PE family. Metallocene PEs with their narrow composition distributions are also excellent candidates for the most demanding applications.

A film COF of 0.2 will meet the demands of most packaging machines. The COF is controlled by the precise amount of slip and antiblock additives in the formulation. It is recommended to source the polyethylenes ready formulated with the additive package and avoid formulating by masterbatch.

Transparency of films depends on resin type, extrusion conditions and cooling rate. Ionomers and mPE provide low haze films from the blown extrusion process.

For box-liners, e.g. cereal boxes, transparency is usually not an issue. Similarly, the coextruded milk pouch has to be opaque to UV light.

A number of laminates and coextrusions used in food packaging are shown in **Table 18.1**. The OTR and MVTR values are included. These illustrate the wide range of barrier properties that can be achieved. The PA-EVOH-PE combination develops the best barrier in the absence of aluminium foil and metallization.

The development of coextrusion techniques that split the melt streams into multiple micro or nanolayers are increasingly practiced. These processes can result in significant property improvements, particularly gas barrier, which can also allow for cost reduction. They are very effective in improving the properties of PA/EVOH compositions. The various film structures discussed below can be improved by using these advanced coextrusion processes.

Table 18.1

Typical plastic films used for food packaging

Structure	OTR	MVTR
PA/EVOH/tie/PE	0.7	1.9
BOPP/met. OPET/PE	1.6	0.5
OPET/PE/EVOH/PE	3.6	4.5
BOPP/PA/PE	8.5	4.2
PA/PE	180	6.0
BOPP/CPP (cast)	900	4.1
BOPP/PE	1280	5.0
OTR: $cm^3/m^2/day$. MVTR $g/m^2/day$.		

Printing is usually by rotogravure or flexographic processes.

18.1 Bag-in-box (BIB) laminates

A bag-in-box (BIB) is a flexible laminate bag, with a dispensing spout or valve, encased in a cardboard box or wooden crate, see **Figure 18.1**. The system is used to dispense liquids and pourable foods and sauces. Typical applications are in the packaging and dispensing of wines, fruit juice beverages, juice concentrates, soups, tomato pastes, edible oil, water, etc. BIB packaging is also used in industrial applications such as battery acid, liquid detergents, motor oils and photo-chemicals. Practically every commercial pourable product can be considered for BIB packaging.

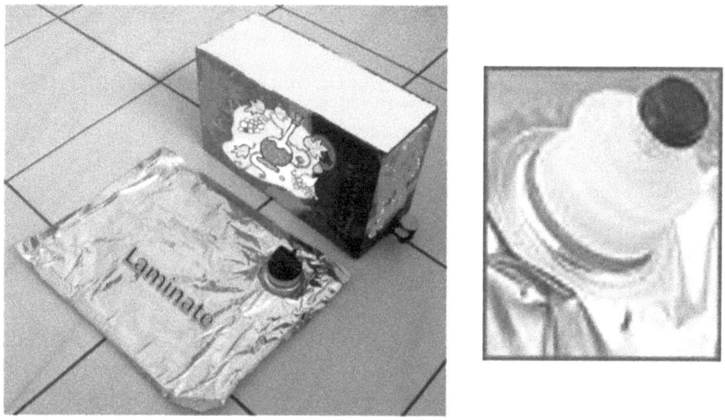

Figure 18.1

Typical bag-in-box system

Aseptic filling where the food and bag are sterilized separately is widely used. The bag is usually a bag within a bag. The bag including the welded-on tap or dispenser is sterilized by gamma (γ) irradiation. The second (outer) bag provides cushioning and protection from leakers. The inner bag is constructed from 50-100 μm PE and the outer bag is usually based on a metallized OPET laminate with a PE type seal layer where gas barrier is specified.

Bag-in box systems can range from small 2 litre sizes up to 2000 litre bulk containers. Popular sizes are 3, 5 and 10 litres. The end-user requires seal integrity combined with flexibility and puncture resistance during handling and transportation. In addition excellent flex crack resistance is required and the package must protect its contents from odour, taint and aroma loss. It is important that the bag is sufficiently flexible and able to collapse as fluid is drawn out through the dispenser.

The advantages of BIB can include the following:

- Longer shelf-life after opening: up to 8-10 weeks
- Low resource usage (10 g/litre) and recyclability
- User-friendly and tamper evident
- Aseptic/non aseptic, hot/cold filling possible
- No cleaning costs
- High brand recognition
- Prevents oxidation in wine packaging (collapsible bag)

The following film properties are critical for bag-in-box applications:

- Sealing (low initial seal temperature, through contamination)
- Hot tack strength at low heat-seal temperatures
- Puncture resistance
- Compliance with food packaging legislation
- ESCR (environmental stress crack resistance)
- Softness and flexibility: bag must collapse on emptying
- Low taint/odour/scalping
- Pinhole-free
- Gas and moisture barrier preventing oxidation and moisture loss

Most of the BIB films are of two-ply construction as mentioned earlier, i.e. bag in a bag, but they can also be of three or more plies. Traditionally the inner ply consists of a tough, high strength seal resin, such as EVA copolymer of 9-12% VA content. Typical thickness is 60-80 μm. However LLDPE, very often blended with EVA, is rapidly penetrating this market because of its better ESCR, toughness, seal and impact strength. The increased cost of higher α-olefin LLDPE or metallocenes can be compensated for by downgauging the film.

Heat-sealable high gas barrier films for BIB applications are based on the following structures:

- Lamination of metallized polyester (OPET) with LLDPE/EVA sealant
- Lamination of PVDC-coated OPET with LLDPE/EVA sealant
- PA film and metallized PA film, with LLLDPE/EVA as sealant
- Coextrusion: LLDPE/EVOH/LLDPE

OPA embodying very high strength is recommended and can be laminated to OPET and PE films. For wine packaging, metallized OPET adhesive laminated with EVA (5-9% VA) or increasingly LLDPE is the most common structure. Metallocene PEs with their advantage of very low extractables and outstanding seal strength are good candidates for this market segment.

Special grades of EVOH with high flex crack resistance are available. These will usually be specified with high ethylene content. The moisture sensitivity of EVOH may limit the wider use of these structures. EVOH can be coextruded with polyamide (type 6.66) without a tie-layer, and this combination has outstanding barrier performance and very low aroma loss. Laminating this structure to a polyethylene-based sealing resin can provide an outstanding film of high strength and flex crack resistance. Metallized films and aluminium foil laminates may crack on flexing and have poor resistance to acid foods (pH < 6.4). Improvements in coextrusion processes enabling the splitting of the PA and/or the EVOH into microlayers can significantly improve both barrier and strength.

The following is a typical laminate structure for an outer ply:

EVA/LL/LDPE 50 μm
met. OPET 12 μm
EVA/LL/LDPE 50 μm

Large liners (1000-2000 L) for beer and wine storage and dispensing will typically have the following structure (the tie-resin will be a maleic anhydride grafted PE for adhesion to EVOH):

Slip sheet (inner ply)	LD/LLDPE bag in contact with liquid
Outer bag (barrier)	Coex. LLDPE/tie-resin/EVOH/tie-resin/EVA

18.2 Liquid pouches

The best example of a successful demanding liquid packaging application based on polyethylene film is the milk pouch, first commercialized in Canada and Switzerland and now also widely used in India. In Canada 60% of fresh milk is sold in pouches.

The initial concept was based on the use of 2-layer coextruded blown film, which produced tough flexible pouches and crucially the two layers eliminated the risk of pinhole formation. The later availability of LLDPE further enhanced properties and in particular heat-seal strength, which reduced the incidence of leakers. This application was one of the first successful uses for LLDPE. LLDPE, particularly octene grades, dramatically reduced leaker rates of the pouch compared to LDPE. Their higher tensile and impact strengths also enabled cost cutting by downgauging.

The polyethylene pouch is a very cost effective ecological method for packaging liquids. A typical pouch weighs 7 g and will hold 1 litre of fluid, i.e. ~0.7% by weight of packed product. The original milk pouch structure was based on two layer coextruded blown film embodying a 50-60 μm white and 20-30 μm black or brown inner layer for UV protection. Over the years variations on the basic design have been developed such as fitting pouring spouts and various stand-up designs. The 2-layer films are being replaced by 3-layer coextrusions to reduce cost.

India is one of the largest users of PE pouches for milk packaging. Initially these were based on 75 μm LDPE film. By the 1980s these were replaced by 65 μm LLDPE blended with LDPE. Downgauging has continued to 55 μm using improved LLDPE blends and packaging is typically carried out at a rate of 50/min. The chilled pasteurized milk is consumed within the day of purchase. Strong leakproof seals are essential to minimize wastage.

The following film properties are typical, based on 40% LLDPE blended with LDPE.

Density (g/cm^3)	0.918 - 0.922
Melt index (dg/min.)	0.8 - 1.5
Slip (COF)	< 0.3

A more sophisticated concept was the development of the stand-up pouch to mimic the metal can or bottle, improving usability. A laminate, typically OPET/PE, is sealed along three edges and the narrower side is gusseted usually with a round mandrel. The gusset allows the pouch to be expanded to form an elliptical bottom, which allows the pouch to stand up. The unsealed end forms the top of the pouch for filling followed by closure and heat-sealing. A zip closure can be fitted below the seal. Pouring spouts can also be welded on for easy delivery of the packed fluid. **Figure 18.2** shows an open plan view of the laminate. The gusset is formed at one seal, which opens up to form the bottom of the pouch. Typical seals of 7 mm width are shown in this sketch.

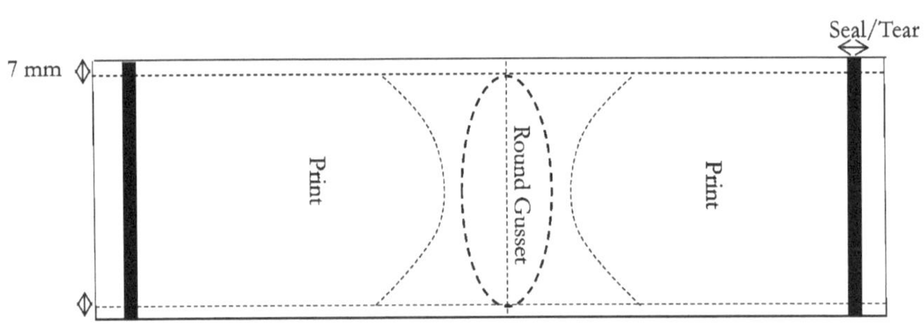

Figure 18.2

Plan view of a stand-up pouch

There are variations in this basic design and seal of the gusset to accommodate different weights and types of product. The plastic pouch is a convenient way to reduce packaging waste, replacing a wide range of products from glass coffee jars to cardboard boxes containing dishwasher detergent. A key requirement for its wider acceptance is the need for resealability and designs that enable pouring without spillage.

Figure 18.3 shows a typical stand up pouch with a resealable zip closure. Closures can be press down or slider zips for resealing. Tear notches and pouring spouts can be added as required. Pouch sizes can vary from 250 ml to 4 litres. A wide range of laminates are used, these can include paper, plastic films and metallized structures. In some applications OPA film is preferred to OPET. The lamination PE film is usually a coextrusion based on LD/LLDPE blends.

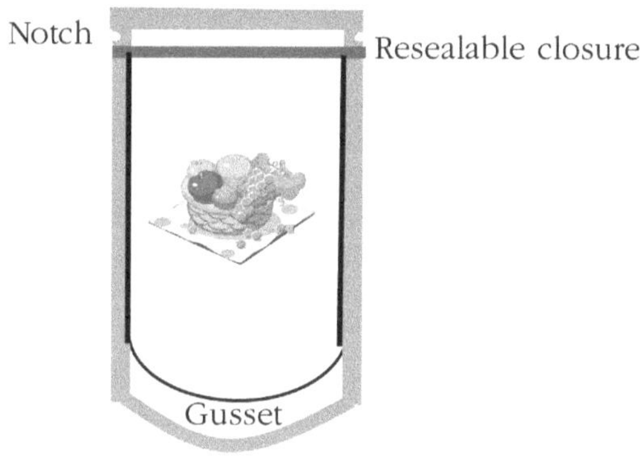

Figure 18.3

A stand-up pouch

There has been some interest in certain markets in replacing PET water bottles with pouches. The aim is resource reduction and lower carbon footprint. These stand up pouches are made from laminates of OPET film with polyethylene. On emptying the pouches collapse and occupy much less space. Because of its blandness water requires an organoleptically neutral package to avoid imparting any off-taste. Selection of the seal layer in contact with the water is critical.

Another innovation is the packaging of wine in pouches. These designs can hold the equivalent of the standard 750 ml glass bottle although larger sizes are possible. The pouch is resealable and can include a pouring spout.

Flexible pouches can be microwavable if required. Designs that include a screw cap welded into the top fold on the pouch are available from specialized suppliers. These can be used for sauces and coffee. In the non-food sector some companies are introducing motor oil and household paint packed in stand up pouches.

Recent announcements claim that stand up pouches made from 100% polyethylene have been produced. These will have the advantage of easier recyclability.

The advantage of pouches is that they can be produced on-line from roll-stock or provided as a pre-made pouch ready for filling and sealing. They can incorporate reclosable zippers, sliders, spouts for dispensing and other features for user convenience.

Some projections indicate that this type of packaging will replace rigid plastic containers in many markets. Resource reduction and lower carbon footprint are the obvious drivers.

18.3 Dry food packaging

Breakfast cereals are one of the most demanding types of dry foods to package. These products are processed and packaged to achieve a shelf-life of one year after production. HDPE is the best moisture barrier. Highly flavoured cereals require additional gas and aroma barriers for protection. Polyamide (nylon) is considered the best aroma barrier.

Film thickness ranges from 38-75 μm. The EVA seal layer accounts for 17-20% of the total film weight. Barrier requirements for cereals are:

MVTR: 3.5 $g/m^2/day$ @ 38 °C and 90% RH

OTR: 100 $cm^3/m^2/day$ @ 23 °C and 0% RH

Some applications may require a MVTR below 2 $g/m^2/day$.

A typical structure will be as follows:

HDPE/HDPE (40 μm)/EVA (8 μm)

Price competition is continually forcing the converters to downgauge by evaluating the most effective moisture barrier materials. The EVA can be replaced by LL/LDPE blends and metallocenes.

Some coextruded structures used as box-liners for cereal and other dried foods are shown in Figure 18.4. The coextruded breakfast cereal box-liner film is typically 35-45 μm in thickness with package volume ranging from 2-3 litres. The last structure in **Figure 18.4** combines HDPE with polyamide (PA6) for moisture and aroma barrier. The tie-resin is a maleic anhydride grafted PE. The polyamide films embody outstanding toughness.

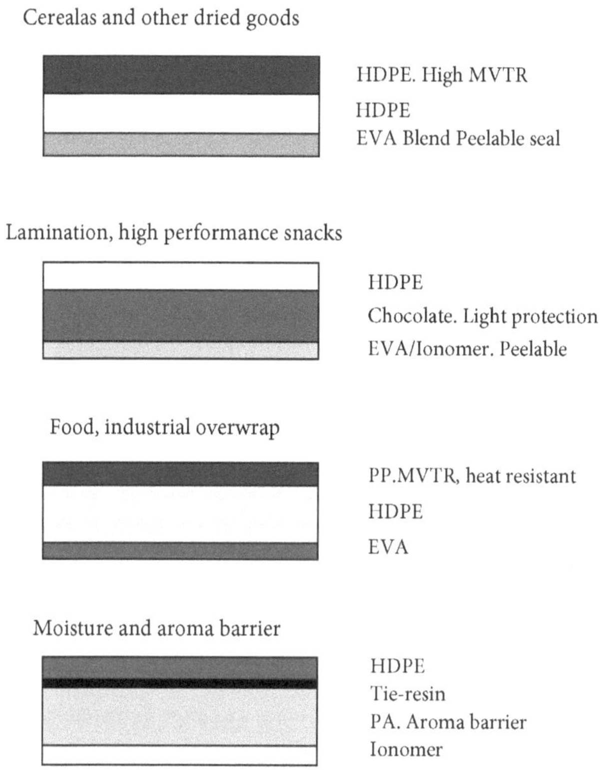

Figure 18.4

Coextruded film structures for FFS packaging of dry goods

Other granular foodstuffs such as sugar are packed in a variety of laminates and coextrusions. The formulations will vary in different parts of the world. Some examples for packaging of rice and sugar granules are shown in **Table 18.2**. OPET, BOPP and in some cases polyamide films are used as the support layer. The OPET and BOPP films can be reverse-printed.

The PE heat-seal layers are various combinations of LD/LLDPE and metallocene PE blends including coextrusions. Strong seals are crucial as well as the ability to seal through powdered contamination. The support film will be reverse-printed to protect the printed information.

Table 18.2

Laminate structures for dry groceries such as rice and sugar

Structure	Package weight
12 μm OPET/60-70 μm LD-LLDPE	1 kg
12 μm OPET/100-120 μm LD-LLDPE	5 kg
18 μm BOPP/60-70 μm LD-LLDPE	1 kg
15 μm PA/150 μm LD-LLDPE	5 kg
55-65 μm PE/ink + adhesive/55-65 μm PE	1 kg
70-80 μm PE/ink + adhesive/70-80 μm PE	5 kg
12 μm OPET/15 μm PA/55-60 μm/PE	5 kg

18.4 Snack foods

Snack foods are typically roasted, fried, salted and flavoured products often consumed between meals. Snacks are highly flavoured and contain fats requiring greaseproof packaging. To retain the texture (crispness) the moisture content must be kept very low. Low MVTR of the package is essential. Shelf stability specifications can range from 2 to 16 weeks and typical examples of these products include the following:

- Potato crisps
- Peanuts, cashews, almonds
- Dried fruit
- Cheese balls
- Nutrition bars
- Wafers

This is a high value-added market where the products can be affected by moisture, oxygen and flavour changes. Fried snacks in particular are spoiled by the presence of oil which will lead to oxidative rancidity. These reactions are accelerated by heat, moisture and air. Snack foods are usually packed on automatic VFFS machines at high filling rates. Where fat content is high flushing with nitrogen is required.

Typical flexible film structures are listed below. BOPP and OPET films are widely used with metallized OPET and BOPP normally included as the barrier layer. The laminates are usually produced by adhesive lamination and the adhesive must be resistant to attack by oils and fats. The polyolefin heat-seal layer will generally range from 35-80 μm in thickness although some applications specify 155 μm seal layers. The seal layers can be blends and/or coextrusions based on LD/LL/HDPE and EVA. Ionomers are also used for seal through contaminated surfaces and have outstanding hot tack. The PE layers shown in the structures below include LL/LDPE, EVA and metallocenes:

- BOPP/PE
- BOPP/OPET/PE
- Metallized OPET/PE
- BOPP/Metallized OPET/PE
- OPET/PE
- OPET/Al Foil 9-12 μm/PE
- OPET/Oxide-coated OPET/PE
- Paper/Metallized BOPP

Other structures are based on a PVDC barrier where the PVDC also provides heat-sealability:

- PVDC-coated OPET
- BOPP/PVDC/BOPP
- PVDC-coated glassine
- PVDC-coated glassine/BOPP

The films must be easily printable as brand identification is crucial in this market.

A typical laminate for potato crisps might have the following structure: 15 μm BOPP/10 μm met. OPET/30 μm LL/LDPE, providing the following level of barrier properties:

OTR: 0.7 cm^3/m^2/day
MVTR: 0.5 g/m^2/day

In some formulations the metallized OPET is replaced by metallized BOPP. The OTR of BOPP films is higher than OPET but MVTR is lower with polypropylene.

18.5 Frozen food films (retail)

Freezing entails reducing temperatures below zero so that microbiological, enzymatic and bacteriological reactions are retarded. In freezing fresh produce, the temperature must pass rapidly through a transition from water to ice so that ice crystals are relatively small and do not disrupt the plant tissue cells.

The produce may be frozen either inside or outside of the package. Modern freezing plants use high velocity cold air or liquid nitrogen (N_2) to remove heat from the bulk unpacked or individually quick frozen (IQF) produce. Cooking vegetables are very often frozen in IQF form. The frozen produce is then packed in coated paperboard cartons or PE pouches. Care must be taken to avoid sublimation of ice, which can lead to damage to the produce described as freezer burn.

Frozen food films have to meet the following requirements:

- Strength/toughness at -20 °C or below
- Outstanding heat-seal strength
- Hot tack for produce packed in FFS machines
- Low COF of film to metal (applicable to FFS converting)
- Stiffness to avoid snagging on mandrels etc. (applicable to FFS converting)
- Clarity if non-pigmented
- Printability (high quality)
- Antifog for some applications
- Food regulation compliance

Vegetable such as green peas, sweetcorn, broccoli, carrots, spinach, cauliflower etc. are packed in PE bags produced from 50-75 μm film. EVA or LLDPE blends are preferred because of their higher impact strength at low temperatures. The EVA will incorporate 2-5% VA and have a fractional melt index. Properties can be improved by the use of higher α-olefin LLDPEs. Coextruded films are now widely used and offer greater flexibility to optimize cost and properties.

The films are either clear or white opaque (TiO_2) and flexographic printed if required. Owing to possible damage by UV freezer lights and the presence of unsightly ice crystals high levels of TiO_2 pigment are blended into the PE via a masterbatch. Some structures are shown in **Figure 18.5**. There are many variants on these generic structures particularly with many coextrusion options.

EVAs can be replaced by the POE materials (<0.91 g/cm^3 density) and higher density mLLDPEs can be used to add stiffness. In most cases the metallocenes can be used in blends with either LDPE or EVA. The mPEs will add improved hot tack, higher tensile and superior low temperature impact strength. The rationale is to reduce cost and improve processability.

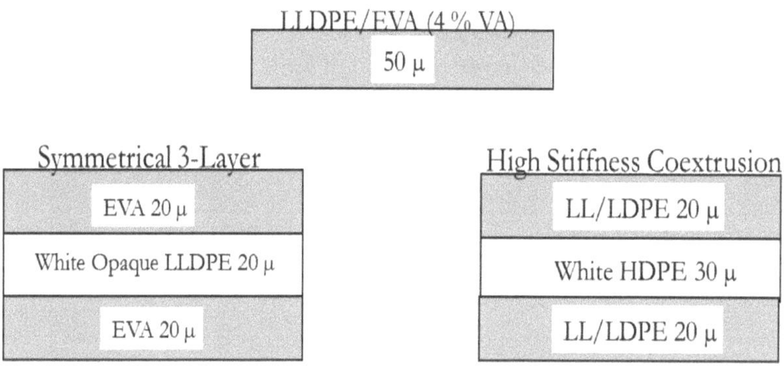

Figure 18.5

Coextruded film structures for frozen foods

Other film structures with improved barrier are listed below:

- OPET/Al/PE
- OPET/PE
- Metallized OPET/PE
- BOPP/Metallized OPET/PE

The PE layer is adhesive laminated or extrusion coated. One layer is normally reverse-printed with logos, nutritional information and promotional materials.

For frozen food films attention must be paid to the strength of the film particularly its seal strength at low temperature. The PE layer is usually coextruded and includes blends of LL/LDPE/EVA and metallocenes to maximize seal strength and minimize cost.

Figure 18.6 compares the impact strength of films at ambient temperature and at -20 °C. The linear resins have a melt index of 1 dg/min. The LLDPE density is 0.920 g/cm^3 and the mLLDPE is 0.916 g/cm^3. The mLLDPE is clearly superior in impact strength. The data shows that the octene LLDPE and EVA are insensitive to the two test temperatures.

Physical property measurements on a 3-layer coextruded structure are shown in **Table 18.3**. The drop tests were carried out at -20°C with filled pouches (500 g). The HDPE layer (MI 0.15/D 0.947) contains 1% TiO_2 added from a PE masterbatch. LD/LLDPE blends were set at 1:1 ratio.

Table 18.3

Properties of 3-layer coextruded films for deep freeze

LL/LDPE blend	HDPE (white)	LL/LDPE blend
15 μm	30 μm	15 μm
LDPE: MI 1.2/D 0.922. LLDPE: MI 1.1/D 0.921		
Property	**Units**	**Value**
Tensile yield MD/TD	MPa	21.6/16.6
Tensile @ break MD/TD	MPa	33.4/28.0
Elongation @ break MD/TD	%	478/862
Seal strength (tested @ 145 °C)	N/15 mm	14.5
Drop test from 1 m @ -20 °C	# of failures from 10	6
Drop test from 2 m @ -20 °C	,,	9

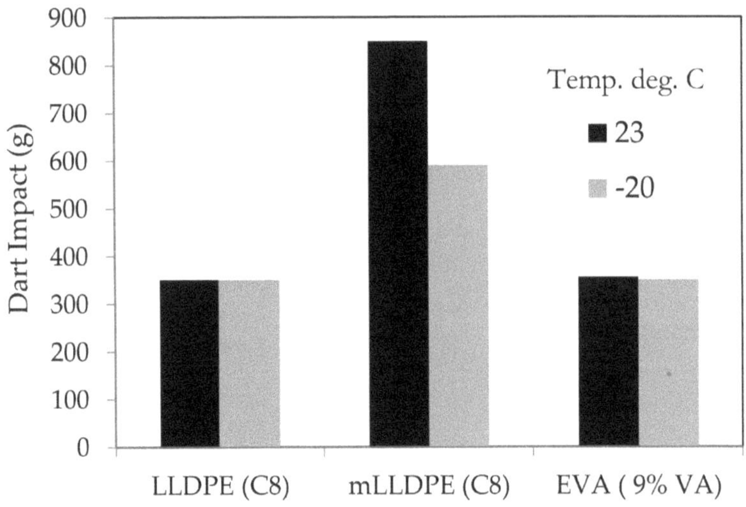

Figure 18.6

Film impact strength at ambient and low temperatures

18.6 Microwave and boil-in-bag films

A great variety of frozen food is packed in flexible pouches and thermoformed packages. The pack is filled with prepared food, evacuated and frozen. Typical examples are fish, rice, pasta dishes etc. The dish is heated in its packaging by the consumer using one of the following cooking methods:

- Microwave
- Boiling in water

The boil-in bag market in the industrialized nations, particularly the USA, has reached maturity. This is mainly owing to other more convenient methods in which to heat prepared foods, such as PET trays, and the limited freezer space in supermarkets. However, boil-in-bag packages do have some advantages:

- Economical light weight flexible pack
- Easier disposability
- Product quality appreciated by some consumers

These bags must meet the following specifications:

- High temperature resistance (87-100 °C during 20-30 minutes)
- Low temperature toughness (-20 to -40 °C)
- Strong heat-seal strength
- Low gas permeation
- Good resistance to greases, fats and moisture
- Good thermoformability and machineability
- No aroma loss or taint

Since PE does not on its own meet some of the requirements listed above, the polymer must be coextruded with or laminated to other materials. LLDPE or MDPE can be laminated to OPET or polyamide to fulfil the requirements of this market segment.

The boil-in bag pouch consists typically of the following basic film structures:

- 2 coextruded or laminated non-formable webs
- 1 coextruded formable web and 1 coextruded non-formable web
- 1 laminated formable web and 1 laminated non-formable web

Films for this application are produced by either blown or cast film processes. Traditionally boil-in-bag films have been laminated, however, the trend has been changing to coextruded and extrusion-coated structures in order to lower costs and eliminate the use of solvent-containing lamination adhesives.

The formable web is usually 75-120 μm thick and will form cavities ranging in depth from 15-25 mm. The structure will be made up of 75-85% LLDPE or MDPE and the balance made from polyamide as the outside layer. The PA layer will be usually 20-25 μm in thickness.

The non-formable film is typically 50-65 μm thick and is a lamination or coextrusion of PA or OPET as the outside layer with LLDPE or MDPE as the sealing layer. The PE accounts for ~80% of the non-formable film. The PA or OPET layer will range in thickness from 12-14 μm.

Some typical examples are summarized in the **Table 18.4**.

Table 18.4
Boil-in-bag structures

Structures	Description
Thermoformed	2-layer film: PA6/LD, LL or MDPE (1) Total thickness: 75-125 μm. 80% PE Thermoformed cavity: 12-25 mm
Pouch	2-layer film: PA6 or OPET/LL or MDPE seal
Pouch	Monolayer for rice: perforated MDPE with proprietary additive (2)
(1) Coextruded with LD/LLDPE-g-MAH tie-layer, or laminated *(2) Development jointly by Uncle Ben's and Tredegar Ind. (USA)*	

Microwavable packages are also produced with PVDC-coated OPET backed with paperboard for reinforcement.

In thermoformed or non-thermoformed structures, LLDPE or MDPE are preferred as the sealant layer. Currently the majority of LLDPE used in these applications is C_4-based. Higher alpha olefins can also be used for improved toughness and heat resistance. PE density ranges from 0.920 to 0.935 g/cm^3 and the melt index from 1.0 to 2.0 dg/min.

18.7 Modified atmosphere packaging (MAP)

The success of MAP is driven by the ability to maintain freshness and quality over an extended time period, e.g. increasing shelf-life from a few days to up to 3 weeks without preservatives. Depending on the produce, boosting shelf-life can involve lowering oxygen (O_2) and replacing with nitrogen (N_2) and carbon dioxide (CO_2) to inhibit bacterial growth. To preserve red meat colour carbon monoxide (CO) can also be used in certain countries.

The modified atmosphere packaging process (MAP) for living plant tissue is based on the reverse photosynthesis that takes place once a living plant is harvested and pre-cut. The respiratory process is reversed with the plant taking up oxygen and exhaling carbon dioxide. MAP must equilibrate the respiratory gases inside the package. The film must allow restricted passage of oxygen (O_2) in and carbon dioxide (CO_2) out, until equilibrium is reached between the respiration of the produce and the diffusion of the gases. The rate at which oxygen is taken up while the carbon dioxide, water, ethanol and aldehydes which cause off-flavours are released is defined as the respiration rate. The plant will also lose water inside the sealed pack forming condensation. The packaging film must, therefore, allow sufficient moisture to escape to minimize mould formation, but the loss must not be excessive and allow the produce to dry out. The process is illustrated in **Figure 18.7**.

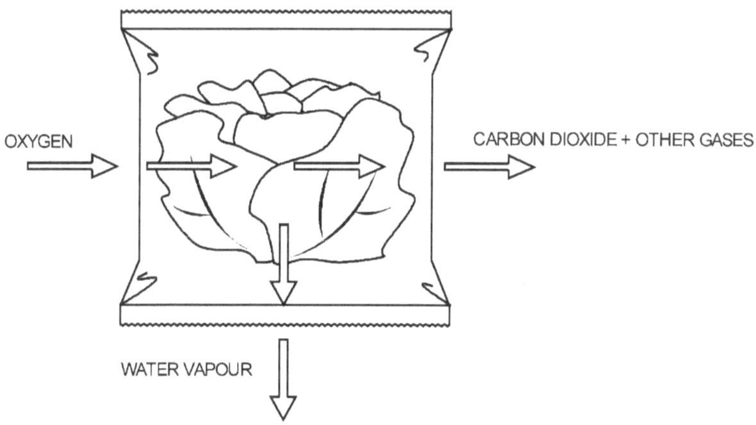

Figure 18.7

The respiratory process in modified atmosphere packages (MAP)

The MAP market is divided into two main sectors; food service and retail.

Food service typically requires a shorter shelf-life and uses large packs (3-12 kg). Aesthetic requirements such as print quality, clarity and gloss are minimal. MAP has caught on rapidly in the food service industry with the realization of cost savings obtained with centralized processing, reduced spoilage and improved product quality.

Smaller packs (0.3 to 2 kg) are used in the retail sector. Attractive aesthetics are required particularly sparkle, low haze and stiffness associated with consumer perception of freshness. Turnover is slower than in the service sector requiring extended shelf-life. Shelf lives of up to 21 days have been reported for certain vegetables.

Different plant tissues have different respiratory rates, which must be taken into account when designing the packaging system. Preparation such as slicing and grating will increase surface area, which accelerates the respiratory rates significantly and must be allowed for in the package design. The ratio of the film surface area to produce weight or volume will also influence the balance of respiratory gases inside the pack. A large 5 kg pack will require a different permeability film to a smaller ½ kg pack for the same produce. For some plants if the film permeability is too high, oxidative degradation will take place and if too low anaerobic browning will occur. Therefore, the film's gas transmission rate (GTR) has to be optimized for each application.

The respiratory rates are determined by measuring the rate of carbon dioxide (CO_2) exhalation per kilogram (kg) of produce. Some values are reported in **Table 18.5** at two temperatures.

Table 18.5

Typical respiratory rates for various foodstuffs

Produce	(mg CO_2/kg @ 10 °C)	(mg CO_2/kg @ 15 °C)
Tomatoes	12-18	16-28
Bananas (green)	5	21-23
Cucumbers	23-29	24-33
Iceberg lettuce	21-40	32-45
Cauliflower	32-36	43-49
Salad	32-46	51-74
Radishes	31-36	70-78
Brussels sprouts	63-84	64-113
Broccoli	75-87	161-186
Mushrooms	100	180

The key film properties to be considered in MAP are:

- Gas transmission rate (GTR) of O_2 and CO_2
- Water transmission rate (MVTR)

Films also require the following physical and optical properties:

- Low haze
- High gloss
- Low seal initiation temperature (SIT)
- High hot tack

- Stiffness for FFS processes
- Strength at low temperature

In designing the total MAP system for a given plant tissue the following parameters have to be considered:

- Product weight (W)
- Respiration rate of produce at storage temperature and gas partial pressure (R_{tp})
- Film thickness (T)
- Package surface area (A)
- Ambient gas partial pressure (P_a)
- Optimum internal gas partial pressure (P_{oi})

The film transmission rate (TR_f) is calculated from equation 18.1.

$$TR_f = \frac{k \times (W \times R_{tp} \times T)}{A \times (P_a - P_{0i})} \quad (18.1)$$

As a guideline it is important to remember that film permeability will change by 4-5% per °C change in ambient temperature.

Unfortunately respiratory rate at actual storage conditions is variable and subject to produce variety, harvesting methods, processing conditions and timing. Equation 18.1 is, however, still useful as it allows calculation of the effects of a change of any of the variables on the internal partial pressure (P_{oi}). By measuring the equilibrium head space gas concentration inside the package with a specified plant packed in a film with a known TR_f the respiration rate (R_{tp}) of the plant can be calculated. Using this number in equation 18.1 with the optimum gas partial pressure will yield the film transmission rate (TR_f) for any given change in pack dimension or filling weight.

Types of polymers for use in MAP applications are shown in **Table 18.6**. The GTR data were determined at 23 °C. Storage conditions for fresh produce will usually be at 3-5 °C, which will slow the respiratory rate and reduce GTR, therefore corrections must be made to allow for actual usage temperatures. For polyolefins the ratio of OTR at 5° and 23 °C is ~3. For CO_2 the TR ratio is ~2.

GTR is clearly increased as PE density is reduced, i.e. comonomer increased. All the PE resins have higher GTR than PP. EVA tends to form sticky films, which can be difficult to convert. The metallocene PE grades have low extractables, are less tacky and have very low haze. However, the films are highly flexible and lack the rigidity of PP. The consumer preference for rigid sparkling films and the greater difficulty of handling extensible films on FFS machines at high speed may limit the use of low density mPE type resins as monofilms. Coextrusion of these mPEs with a more rigid material presents an obvious solution to tailor the properties for this demanding application.

Table 18.6

Comparative gas transmission rates (23 °C)

Polymer type	Composition/ density (g/cm^3)	OTR (25 μm) ($cm^3/m^2/day$)	MVTR (25μm) ($g/m^2/day$)	CO_2 TR (25 μm) ($cm^3/m^2/day$)
PP	Homopolymer	3500	5	14000
LLDPE	0.920	7000	10	31000
EVA	9%VA	11000	70	80000
mPE (hexene)	0.900	12750	15	62460
mPE (hexene)	0.910	7500	19	33540
Terpolymer m-PE	0.900	14200	25	75300
PVC (plasticized)	-	11870	473	-

The retail sector has traditionally used cast PP/PPRC random copolymer coextruded films for MAP of fresh produce. PP films are low cost, have low haze, high gloss, high stiffness and adequate strength and seal performance for most FFS processes. However, the relatively low gas transmission rates (GTR) of PP films has limited shelf-life to 3-5 days with most produce. The low OTR of PP films will lead gradually to anaerobic conditions and spoilage caused by oxygen extinction as the respiratory process converts the oxygen into exhaled carbon dioxide. High GTR films were, therefore, evaluated over the years with the aim of further extending shelf stability by better control of the respiratory process. With coextrusion and/or lamination it is now possible to control the GTR of the film by the choice and thickness of polymers in the structure. From a theoretical basis as shown in equation 18.1, combined with field experience, converters have been able to devise empirical relationships which correlate a plant's respiratory rate with the design and size of the package in order to maximize its shelf-life. Other properties such as toughness, hot tack, SIT and clarity can also be optimized by the constituents of the coextruded film. For applications requiring very high quality graphics, a reverse flexographic printed film may be adhesive laminated to the heat-seal film, typically LLDPE, EVA or mPE.

Some permeability data for an LLDPE and mLLDPE (both octene) coextrusion published by Dow Chemical is shown in **Table 18.7**. The coextruded structure is configured as follows:

LLDPE (octene)	15%
mLLDPE (D: 0.902 g/cm^3)	70%
LLDPE (octene)	15%

The very low density metallocene contains 12.5% octene comonomer (density: 0.902 g/cm^3) and has a melt index of 3.0 dg/min. The film was extruded by a cast coextrusion process, which partly explains the higher than expected OTR value.

Table 18.7
Permeability of mPE and PVC films

	Coex. mLLDPE	PVC (Plast.)
OTR (cm^3/m^2/day)	22,800	11,870
MVTR (g/m^2/day)	46	473
Elastic recovery (%)	88.0	89.6

Compared to plasticized PVC the polyolefin film has double the OTR and 1/10 the MVTR.

The technology is increasingly being used beyond the retail sector in agricultural production. In developing countries where transportation is slow and refrigeration not available, polyethylene MAP bags are being used to preserve produce for export for up to one month, significantly reducing the amount of spoilage. Typical bag sizes are from ½ kg to 20 kg. These are claimed to control the permeation of oxygen, carbon dioxide and ethylene gas.

As a final comment in addition to the selection of the film some packs may include additional modifiers in the form of desiccants and ethylene absorbers to control the ripening processes. This technology is evolving very rapidly as the demand for fresh produce of the highest quality increases.

18.8 Bakery films

The foodstuffs included in this segment are bread, rolls, pastries, cakes, pies, crackers and biscuits. The films need to deliver toughness, puncture resistance, good optics and reliability. With the current use of freezers the packaging also needs to be suitable for long term use at sub-zero temperatures.

Specialty films used by this sector are single layers and composite structures based on BOPP, cellophane and HDPE. BOPP and HDPE are making rapid inroads into the cookies and crackers markets by replacing cellophane and glassine (paper). Multilayered films are produced by lamination or coextrusion. Some mono and multilayered films are listed in **Table 18.8**.

The BOPP films offer good barrier properties particularly when combined with a PVDC coating. They are used on form, fill and seal or overwrap packaging machines. Coextruded HDPE/EVA films are used on VFFS and HFFS lines to package crackers and cereals. A peelable seal layer is preferred.

LD/LLDPE films are converted into bags by side-weld sealing. These are used to package sliced bread and soft products such as muffins and buns.

Some bakery products require breathable films which are produced with proprietary technologies. These involve special laser perforation techniques or the use of compounded inorganic additives. PP is normally used in applications requiring breathable films.

Table 18.8

Films for bakery products

Polymer	Film type	Thickness (μm)	Application
PP	BOPP	12-20	Overwrap film
Coextruded	PPRC/PP/PPRC*	17-35	Bags for bakeries, tray overwrap
	PVDC/BOPP/PVDC	16-18	Biscuit packs
PP	Perforated film	-	Bags for fresh bread, e.g. store-baked baguettes
PP	CPP cast film	25-30	Biscuit bag liner
Coextruded	PP/PE/EVA	25-75	Snack cakes
HDPE	HDPE/EVA HDPE/LDPE/EVA coextruded	20-50	Crackers, biscuits
LDPE	Usually blown film (often white)	25-50	Bread bags and buns

**Propylene-ethylene random copolymer.*

18.9 Fresh meat, cheese and fish produce packaging

Table 18.9 shows a list of films used to package meat products. PVC and EVA cling film is used to wrap individual meat cuts placed on trays at point of sale kept under chilled conditions. These are non-barrier films and provide limited shelf stability. The use of mLLDPE to improve heat-seal has been evaluated in several applications. The multilayered films are produced either by coextrusion or lamination.

Table 18.9

Materials used in meat packaging

Monofilm	**Key function**
PVC (plasticized)	Cling film for meat overwrap
EVA	Cling, high flexibility
EVA/LLDPE	Non-barrier shrink bags for frozen poultry
Multilayered	
EVA, LD/LLDPE, Ionomer	Shrink, heat-seal, low cost
PA	Barrier, strength and drawability
OPET	Clarity, strength usually as lidding (top web)
EVOH	Oxygen barrier
PVDC-coated or coextruded	OTR, MVTR barrier
Mod. polyolefins (PO)	Tie-resins in coextrusion

Frozen poultry (turkeys, chicken etc.) are traditionally packed in EVA or LLDPE shrink bags. The EVA (5% VA) and LLDPE films are preferred over LDPE because of their superior impact strength at low temperatures. Blends are also used to optimize cost/performance ratio. Vacuum processes such as the Cryovac® process based on irradiated coextruded EVA/PVDC/EVA films are used where outstanding toughness and balanced shrink are specified. The double bubble process ensures balanced shrink of the irradiated film.

Red meat is often vacuum-packed for distribution in barrier shrink bags or pouches at central processing plants. Typical film structures are:

- Shrink film, e.g. coextruded EVA/PVDC/EVA (Cryovac® irradiated bi-oriented film is market leader)
- Non-shrink film, e.g. PA/ionomer or EVA (EVOH or PVDC is included by some suppliers)

An OTR of 30-40 $cm^3/m^2/day$ @ 23 °C and 0% RH is usually acceptable if the product is refrigerated. The shrink bags are hand-filled, evacuated and clipped (closed) and then conveyed to a hot water bath where the bag shrinks to form a skin tight layer around the meat. The operation is carried out on automatic machines. EVA (9-12% VA) has excellent shrink properties at temperatures just below that of boiling water. When irradiated by electron beam (Cryovac® process) to induce some crosslinking and then reheated and biaxially oriented in a bubble process, the film combines outstanding toughness with high and balanced shrink tension. The Cryovac® bags are available in a thickness range of 60-90 μm.

For the packaging of processed meats (e.g. salami) and some sliced cheeses, thermoformed trays are preferred. Typical structures for the lidding and the bottom films are shown below. The PE films are typically blends. Tie-layers are required to bond the PE with PA and EVOH for the bottom web.

LIDS:	PVDC/OPET/PE
BOTTOM FILM:	PE/PA **or** LLDPE/EVOH/LLDPE/PA

Removable lids for aluminium trays (typically used on products to be placed in an oven) will be OPET/OPET peelable.

18.10 Film for coffee packaging

Coffee packaging is a very demanding application and requires high performance laminates.

The following properties are important in coffee packaging:

- Puncture resistance
- Flex crack resistance
- Dead fold
- Seal through powder contamination
- Low extractables
- No flavour loss
- No static

In vacuum packs outstanding gas barrier is essential. However roasted coffee generates carbon dioxide at a rate which depends on the granule size. The finer the granules the faster the degassing rate. Hence there are patented systems such as one-way valves and carbon dioxide absorbent compounds which can be introduced into the pack. The key properties are provided by different components of the laminate as shown in the typical generic structure in **Figure 18.8**.

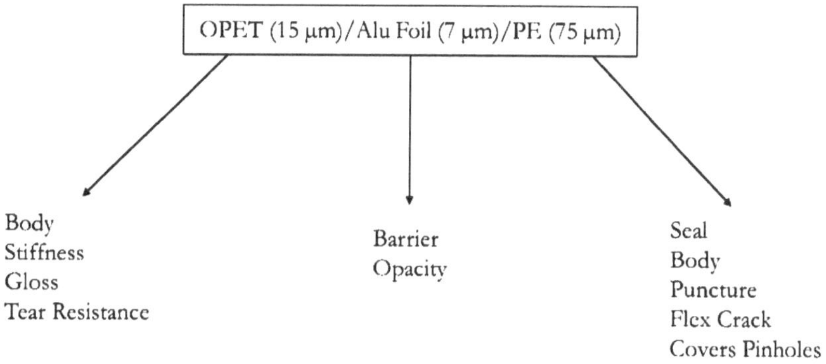

Figure 18.8

Generic structure for coffee packs

A variety of other laminate structures have also been developed for this demanding application, traditionally based on PE (blends of LL/LDPE and/or coextrusions) laminated to PVDC, foil or a metallized polymer layer, plus an oriented polymer or paper outer layer.

However, aluminium foil is generally being replaced by metallized films, and the driving forces are both environmental and cost. Additionally the replacement of OPET/PVDC by OPET/SiO_x films is offered as a means of eliminating the use of the chlorinated polymer, and although there are still some paper/PVDC-coated materials in use these are being phased out.

Another innovation is the replacement of the traditional barrier layer with EVOH. A 5-layer coextrusion of LLDPE/tie-layer/EVOH/tie-layer/LLDPE laminated to OPET is one solution. EVOH-based laminates for vacuum coffee pouches have become successful in Japan and more recently in Scandinavia and Switzerland for environmental reasons. One of the latest concepts for coffee packaging is shown below where EVOH barrier and metallization are combined and laminated to reverse-printed BOPP. The sealability and puncture resistance of the 3-layer PE film can be improved by blending with 20-30% mLLDPE. It is not clear whether the extra cost of combining metallized OPET with coextruded EVOH is likely to be cost effective.

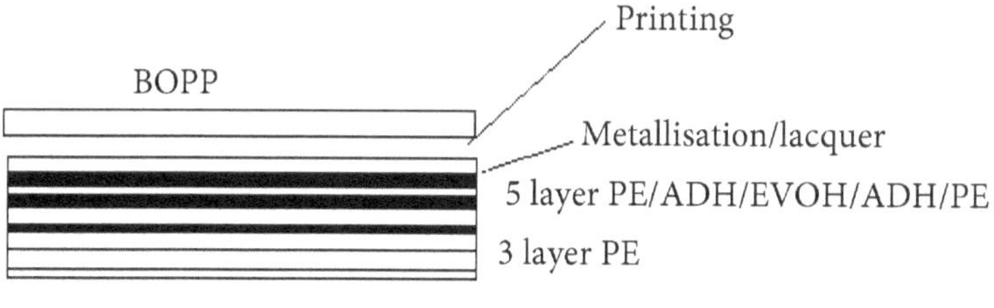

The two principle markets that utilize flexible films for coffee packaging are institutional packs, e.g. coffee machines in offices, and vacuum packed coffee for home use.

Institutional packs are usually 2-ply films based on 12 μm metallized OPET laminated or coated with 50 μm LDPE. The OPET film is printed and over-lacquered if necessary for abrasion resistance. For larger sized packs (2-3 kg) the higher strength metallized BOPA (biaxially oriented PA) is preferred over OPET and the following structure is typical:

15 μm metallized BOPA/PVDC coating/60 μm LDPE or LLDPE.

This film offers greater puncture and flex crack resistance then OPET, but carries a premium price.

Some typical structures of vacuum packs for home use are as follows:

1. 12 μm OPET/8-9 μm Al foil/15 μm BOPA/75 μm LD or LLDPE
2. 12 μm OPET/20 μm BOPP/8-9 μm Al foil/85 μm LDPE
3. 12 μm OPET/8-9 μm Al foil/85 μm LDPE

Film #1 is the most popular as it provides high quality printing on the OPET and outstanding puncture and stress crack resistance from the oriented polyamide film. LLDPE provides superior seal strength and hot tack. Film #2 provides a 12 month shelf-life for ground coffee. Structure #3 with just 3 layers has a shelf-life of 6-8 months and is the least popular film.

18.11 Retort packaging

The retort process was developed for sterile packaging of a variety of edible products ranging from water to fully cooked meals in a flexible pouch. The technique is a convenient lightweight alternative to metal cans. The pouch is constructed from a plastic laminate which must withstand temperatures of 116-121 °C. The food is cooked for several minutes under high pressure inside a retort or autoclave where all micro-organisms are killed. No permeation of gases is allowed. The process was originally developed for military rations and later developed for ready meals and other applications such as cooking sauces. Pet foods are a large market for retort packages. The films recommended for the construction of the laminate are generally based on the following materials:

- OPET: provides a glossy rigid layer, which may be reverse-printed

- BOPA: provides puncture resistance
- Aluminium foil: provides gas and moisture barrier
- Cast PP (CPP) film: heat-seal layer

A PP copolymer (PPRC) is normally used for the seal layer, however some processors have claimed that homopolymer PP is preferred because of its higher heat resistance.

Adhesives and printing inks must be resistant to the retorting process. The adhesive formulation must also be approved by the appropriate regulatory agency and completely cured. These are usually limited to aliphatic formulations. No migration of residual trace materials or blistering is allowed. Some converters will condition the laminate in warm environments (40-45 °C) for several days to complete the cure. Pouch sizes may vary from 100 g to 3 kg portions. A typical laminate will have the following structure:

12 μm OPET/9-15 μm aluminium foil/76 μm CPP

A polyamide (PA) layer can also be included between the foil and the CPP for added strength.

Since aluminium is not suited for microwaving this has led to an interest in alternative barrier films. The best candidates are oxide (SiO_x, Al_xO_y) coated films of OPET, OPA or BOPP. The best results in terms of barrier use oxide-coated OPA and OPET, with BOPP the least satisfactory. Whereas with aluminium foil diffusion occurs through cracks and pinholes, in oxide-coated films diffusion occurs through nanoscale defects in the oxide layer. This can lead to loss of barrier during the high temperature retorting process as well as during storage, and this may limit the shelf-life and type of food product packed. EVOH barrier layers have limited use in this application owing to their high moisture sensitivity and loss of barrier at retorting temperatures. However, there are claims that some EVOH producers have been developing specialized grades for retorting. Another issue when not using Al foil is the avoidance of stress whitening during retorting.

In developed markets where refrigeration is ubiquitous, there is less demand for long shelf-life. Other techniques, such as high pressure pasteurization or "pascalization" are used although these are not suited to all foods. The process works at low temperature using pressures of approximately 400 MPa, which kills many micro-organisms, and the treatment time can be as low as 2-5 s. EVOH laminates can be considered for this process since there is no high temperature requirement, and PE-based seal layers may also be used.

18.12 Rice and pasta

Dry rice and pasta products are packed in pouches that require moisture protection. Oxygen barrier is not a critical property. The film needs to resist penetration by the hard and sharp rice and pasta products during shipping and handling. Three examples of films are:

- PE/PE laminate
- Paper/CPP
- BOPP/seal

The PE/PE laminate is normally based on blends. One side is reverse-printed with the logo etc. A typical blend might be based on the following constituents:

mLLDPE-C_6,	MI= 1.0 dg/min. 40%
LLDPE-C_4,	MI= 1.0 dg/min. 40%
LDPE,	MI= 2.0 dg/min. 20%

The tough metallocene LLDPE maximizes puncture resistance. The use of butene LLDPE will reduce cost and the LDPE will avoid the formation of sharkskin during extrusion. There are some variations that use EVA in the blend and octene LLDPE can also be used. The total laminate is 60-70 μm in thickness.

Chapter 19

Agricultural films

A very large market sector for polyethylene films is in agriculture and horticulture for crop protection. Plastic film offers less protection than glass in colder climates, but more in warmer climates for loss of heat and water. Related applications include polyethylene film for mulching, seed conservation, disease and pest control and conservation. It is estimated that on a global basis some 5-6 million hectares of land is covered with plastic, mainly polyethylene film. The key applications are segmented as follows:

- Mulch, cover, hydroponic films
- Greenhouse and tunnel film
- Bale silage stretch film
- Soil fumigation and solarization film
- Pond liners/geomembranes (<2 mm)

The types of PEs used in agricultural film applications are listed in **Table 19.1**.

The addition of 25% LLDPE to LDPE will improve puncture resistance for tunnel, mulch and silage films. Blends with more than 25% LLDPE can suffer from bubble instability with the large diameter dies and wide layflat widths used in these applications. Coextrusion is increasingly used to enable downgauging and reduce costs. New generation mLLDPEs are used increasingly in these coextruded films to improve tensile strength and toughness.

A wide range of masterbatches are available to meet all the formulating needs of the converter. It is important to use products which can guarantee high quality dispersion of the additives in the final film. Perfect dispersion is required with light stabilizers and applications requiring maximum opacity. It is important to ensure that extruders include well proven mixers in the screw design.

Table 19.1

Polymers for agricultural films

PE Type	MI (dg/min)	Density (g/cm³)	Comonomer	Usage
LDPE	0.2	0.919	-	Mulch, silage, tunnels
LDPE	0.3	0.922	-	Tunnels, silage
LDPE	0.3	0.923	(LS* stabilized)	Greenhouse
LDPE	0.4	0.950	(LS* stabilized)	Greenhouse
LDPE	0.4	0.924	(LS* stabilized)	2 and 4 season film**
LDPE	0.4	0.946	(LS + energy screen)	3 season
LDPE	0.6	0.926		Thermic (IR) tunnels
LDPE	2.0	0.918	-	General purpose/mulch/ cover
EVA	0.3	0.935	14% VA	Greenhouse/tunnels
EVA	0.35	0.935	12.5% VA	Greenhouse/thermic film
EVA	0.6	0.970	5% VA	Greenhouse
EVA	0.8	0.926	6% VA	Thermic film for tunnels
EVA	0.8	0.936	14% VA	LS* and IR film
EnBA	0.8	0.930	8% BA	Tunnels
LLDPE	0.5	0.918	C_4	Silage bags and blending
LLDPE	0.7	0.925	C_4	Silage bags and blending
LLDPE	0.9	0.923	C_8	Blending/tunnels
LLDPE	0.9	0.917	C_6	Tunnels, mulch
LLDPE	1.0	0.918	C_4	Silage bags and blending
LLDPE	1.0	0.926	C_6	Silage bags and blending

**LS: light (UV) stabilized.*
*** Will last more than one growing season.*

19.1 Mulch, cover and hydroponic films

- Mulch and cover films are applied to protect and accelerate plant growth
- Films suppress weeds and conserve water
- Clear and dark films intercept sunlight and warm the soil
- White films reflect heat and reduce soil temperature
- Hydroponics (soil-less cultures) are covered with energy films (see Section 19.2)

LD/LLDPE films are normally used in these applications in linear lean blends. Typical applications include the following:

- Melons: LDPE and LLDPE of 30-35 μm thickness and 1.4 m width films are used
- Cotton: 12 μm LLDPE film of 70 cm width is increasingly being used
- Asparagus: mulch films are 50-62 μm thick and made form LDPE and EVA (8%)

The worldwide usage of mulching techniques continues to show very rapid growth. The technique consists of creating a microclimate around each plant. The benefits derived are:

- Reduced water loss
- Improving the quality of the harvest
- Advancing the harvest
- Improving soil fertility and structure

These films are based on LLDPE and LDPE for standard applications and pigmented (black) when required. Black mulch films must have a light transmission level of less than 1%. Thickness can range from 12-50 μm. Some processors use EVA (5-6% VA) films with thermal-insulating additives and antifog. Photodegradable films are made from LLDPE or LL/LDPE blends with the appropriate photodegradable masterbatch. The heat sensitivity of the photodegradable agent during the extrusion process requires good temperature control.

Black film is used for produce such as strawberries and melons. Transparent films are used with maize and cotton. White asparagus requires a black film to reduce discolouration of the crop caused by light, as well as suppressing weed growth.

Floating films, i.e. films laid on top of the plants, are either perforated or made from non-woven materials.

Black masterbatches are usually added to films used in silage and mulch applications to provide opacity, weathering resistance and the necessary physical properties.

In Southern European countries light transmission levels for mulch films will usually be specified as <1% whereas in many countries specifications are based not on key performance requirements but on carbon black concentration (2-2.5%). However, the type of black, additives, quality of dispersion and dilution will have an important effect on weathering resistance, physical properties and covering power. It is important, therefore, to carefully select an optimum masterbatch formulation for the best results at the lowest cost.

Another class of mulch films makes use of the fact that insects can be selectively attracted by certain colours. The yellow spectrum of this mulch corresponds to the colour co-ordinates of the flowers of certain vegetables. The film is modified with perfumes that attract specific insects. Insecticides are added, as controlled release in the outer layer or applied to the film surface via a coating system. Very often the films are sprayed with adhesives to finish the job. The purpose of the adhesive is not to stick the insect to the film, but to seal the tracheae through which the insects breathe.

Research has also been reported on many other investigations of coloured and silvered mulch films, designed for their beneficial effects on the growth of specific plant types as well as temperature, moisture and weed control.

19.2 Antifog energy screens

Antifog energy screens are films positioned above the plants in a heated greenhouse to create a glazing effect or to develop a microclimate better adapted for soil-less (hydroponic) cultivation.

Unlike packaging films a wider choice of surfactants are available to modify a polymer's wetting ability for condensation. However, if additives that bloom out of the film are transferred to the plant by dripping condensation then there is a risk of phytotoxicity problems. A higher hydrophilic and lipophylic balance of the surfactant improves the antifog activity but will increase the risk of plant damage. To overcome this, complex mixtures of tensio-active substances designed in combination with the polarity of the polymer system to regulate controlled release of the additives are used.

It is not possible to combine all the ideal film properties in a single-layer structure. One of the problems encountered is severe blocking after storage. A second problem is the inability to control migration over longer periods of time, whilst providing a fast antifog activity straight after extrusion. For this reason the majority of the antifog energy screens on the market today are coextruded films.

The outer layer can be a polyethylene of slightly higher density than the other layers. The central layer is a copolymer that can contain large amounts - close to 5% - of the surfactant system. The inner layer will have a lower comonomer content and a much lower concentration of the additive system to create a fast start-up. Usually, after some time antifog is found on both surfaces. However, in harsh climatic conditions (freezing temperatures) only the correct film side will show antifog activity.

The latest developments are breathable films having MVTR of more than 5000 g/24 hr.atm./m^2. These films combine the advantages of woven screens with those of the polyolefin-based films and are used for climate control over special crops.

19.3 Greenhouse and tunnel films

The use of flexible (plastic) greenhouse structures and large tunnels is very advanced in many agricultural regions of the world.

This is a specialized market, which requires substantial investment in large extrusion plant capable of producing very wide films designed to be anchored over a frame as shown in **Figure 19.1**. The majority of agricultural films are produced by the blown film process. Dies of 2.5 m diameter are not unusual in this business. Such equipment is costly to install and operate efficiently. The complexity and investment is considerably magnified where coextrusion is specified. Balanced strength in the MD and TD is needed.

The following film properties are important:

- High strength in both directions
- High impact strength
- High puncture resistance
- High tear resistance

Greenhouse and tunnel films range from 4.5 m to 14 m in width. Film thickness is 180-200 μm.

Profiles for greenhouse and tunnel structures are sketched in **Figure 19.2**.

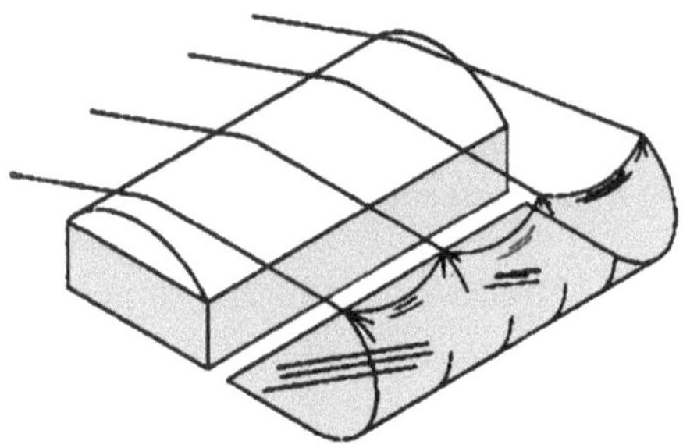

Figure 19.1

Greenhouse and tunnel construction

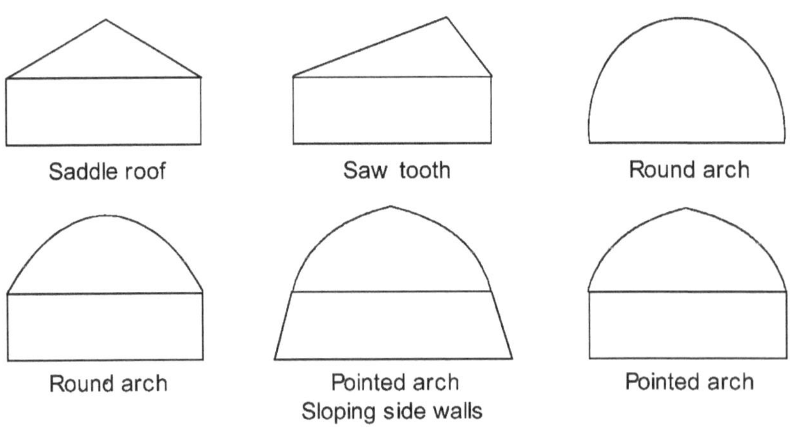

Figure 19.2

Different greenhouse and tunnel shapes

Films for greenhouse applications should meet the energy balance criteria as shown in **Figure 19.3**.

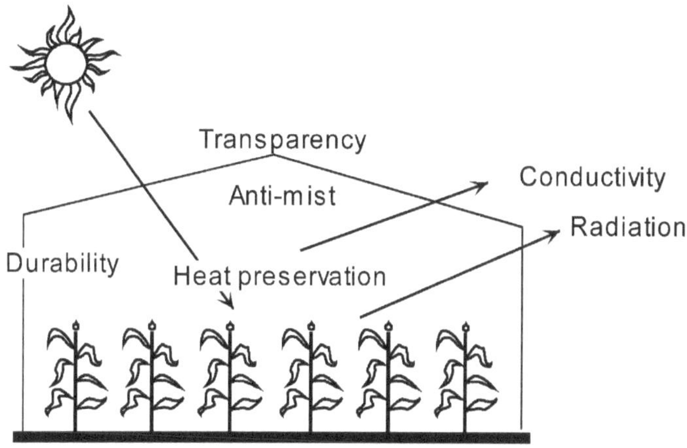

Figure 19.3

Energy balance criteria for greenhouses

The following application segments have shown the highest growth for greenhouse usage in recent years:

- Vegetable crops, e.g. green beans, tomato, pepper, marrow
- Flower growing such as roses, carnations etc.

Critical properties are listed below:

- High visible light transmission
- Low transmission of long wave radiation (3000 to 15 000 Nm)
- No condensation or droplets on the inner surface (fogging) although water as a continuous film is acceptable
- Mechanical strength in both directions
- Low creep
- Impact strength
- Puncture resistance
- Tear resistance
- Resistance to UV light

Resistance to UV light is crucial if the film is to resist at least three years of exposure. Hindered amine light stabilizers (HALS) are formulated in the polymer for UV protection. Exposed to light free radicals are formed which gradually degrade the polymer. The HALS stops the free radical propagation and protects the polyethylene film from rapid degradation. Correctly formulated polyethylene films can be guaranteed for three years of service. Nickel quenchers are also used as stabilizers in PE. These are more costly than HALS, but do provide better protection from sulphur-containing acid rain.

Thermic films are based on EVA and mineral additives and formulated to reduce the loss of IR radiation in the 7-14 μm wavelength. These films will protect crops from freezing at night and will also diffuse light and reduce the effect of shaded areas within the greenhouse during the day.

The IR transmission of LDPE and a range of EVAs are shown in **Table 19.2.** The data shows the reduction in IR transmission with increasing vinyl acetate content. At 150 μm the 14% VA resin has almost one third of the heat loss of a typical LDPE.

Table 19.2
IR Transmission (7-14 μm)

Polymer	50 μm	100 μm	150 μm	200 μm
LDPE	81	69	61	56
3% VA	70	54	44	37
6% VA	64	45	36	29
9% VA	58	40	30	23
12% VA	54	35	26	20
14% VA	51	32	24	18

Coextrusion is the favoured technique for high performing greenhouse films. A typical structure is:

LDPE (25%)
EVA (50%)
LDPE (25%)

Coextrusion makes it possible to combine the attributes of EVA and LDPE i.e. toughness, UV resistance, non-stick surface and low creep in a single film. EVA copolymer plus the appropriate light stabilizer formulation can survive three seasons in the most demanding climatic conditions. Thickness for 3 season film is 200 μm. Some suppliers claim to guarantee 4 season films. In these films the inner EVA layer is a 12-16% VA copolymer of low melt index. The outer layers are also increasingly being produced from 2-3% VA copolymer. Some asymmetrical structures are also produced. These have an outer LDPE layer, a core layer of 14% VA copolymer and an inner layer of 4-5% VA resin.

Greenhouse film applications cannot tolerate the relatively high cold flow (creep) of LLDPE. Blown film dies up to 2.5 m diameter are used to produce the wide film widths and this requires high melt strength polymers to stabilize the bubble. Therefore, fractional melt index LDPE with its high melt strength is predominantly used. In the less demanding tunnel applications, blends with up to 20% LLDPE are used. Octene LLDPE and metallocene PEs will provide improved tear resistance. To reduce cost butene-LLDPE is used in lean blends with LDPE.

The light stabilizers (UV) can be provided by the manufacturer as ready formulated grades or by masterbatch addition at the extruder. Excellent additive dispersion is essential to maximize its effectiveness. Other additives such as IR reflection and antifog reagents may also be needed. Masterbatches are available from specialized compounders. The light stabilizers are of the HALS (hindered amine) type or quenched nickel complexes. However since the International Agency for Cancer Research has classified nickel compounds as carcinogenic products (Class 1), the use of HALS stabilizers is preferred.

The effectiveness of light stabilizers is strongly influenced by the base resin, e.g. the presence of comonomers (VA) and catalyst residues. The regional climate will also play an important role on the resistance of the film.

The results of artificially accelerated ageing tests do not necessarily correlate well with the results of natural ageing. **Table 19.3** shows the improved performance of an EVA tunnel film compared to LDPE in harvesting lettuce.

Table19.3

Lettuce yields under small tunnels

Film	Yield @ harvest	Average weight
LDPE	64%	220 g
12% VA	100%	383 g

Figure 19.4 shows an advanced 5-layer greenhouse film, which is designed to balance the required specifications at the lowest cost. This demonstrates the range of properties required for effective performance.

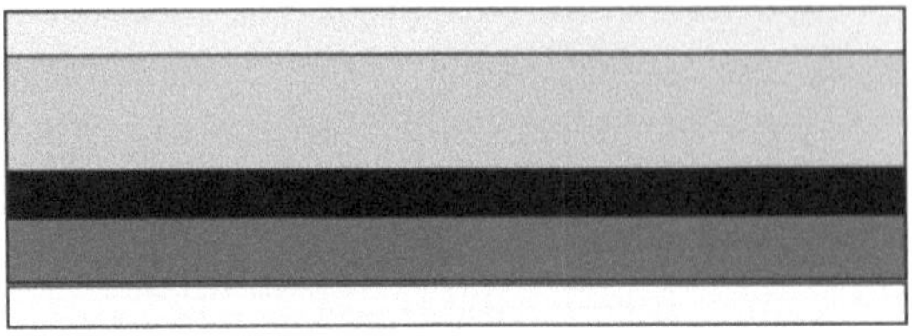

Figure 19.4

5-layer coextruded greenhouse film

19.4 Silage films

Silage bags are generally used to preserve maize under anaerobic fermentation conditions, i.e. fermentation in the absence of oxygen. The bags are also used for preserving hay and grass. The majority of silage films are pigmented with carbon black although some films are white pigmented. The presence of finely dispersed carbon black builds in the necessary protection against UV light degradation. Coextruded films with black on the one side and white (TiO_2) on the other are also available. Coextrusion provides the opportunity to downgauge without loss of strength.

The key requirements for films in silage applications are:

- Puncture resistance
- Tear resistance
- Resistance to climatic stresses (rain, hail, wind)
- Resistance to attack by rodents, farmyard animals etc
- Weather resistance (UV light)
- Low oxygen permeability (anaerobic conditions)
- No pinholes

Classic silage usage utilizes highly pigmented LDPE and LLDPE blended films in a thickness range from 125 to 200 μm. Some properties for LDPE silage films are shown in **Table 19.4**. The bags are coextruded, usually 2-layer black/white films (3-layer in some cases).

Historically the black films were produced from recycled PE. Recently there has been a change in emphasis, which reduced the use of scrap in order to meet stringent performance norms established by the industry. The initial idea was to discourage unscrupulous operators flooding the market with substandard films, however the current situation is to maximize the use of recyclates in order to meet the recycling requirements implicit in environmental legislation. This has led to an increasing use of coextruded films as a means of maximizing scrap content at minimum total thickness. As an example a 3-layer coextruded film will combine outer layers based on high strength LD/LLDPE blends with a core layer made from scrap. The type of LLDPE used and its blend ratio with LDPE are determined by cost at the time of manufacture. Substituting LLDPE with mLLDPE may add sufficient reinforcement to the film and assist in increasing the amount of scrap in the core layer.

Table 19.4

Average properties of coextruded black/white LDPE silage film

Property	Unit	Average value
Average thickness	μm	197
Thickness variation	%	±8
Tensile strength MD/TD	N/mm²	19.5/20
Elongation MD/TD	%	460/790
Dart drop	g	490
Dart drop	g/μm	2.4

Coextrusion dies of 800-2000 mm diameter fed from 75 to 150 mm extruders are used in order to be competitive in this market. The nip rolls should be designed to handle at least 5½ m wide layflat films. If the machine is also used for large greenhouse films wider films will need to be accommodated. The converter will probably need to have substantial outlets in the large tunnel film and geomembrane markets to justify this large investment in machinery.

Bags for grass and hay will generally have the dimensions shown in **Figure 19.5**.

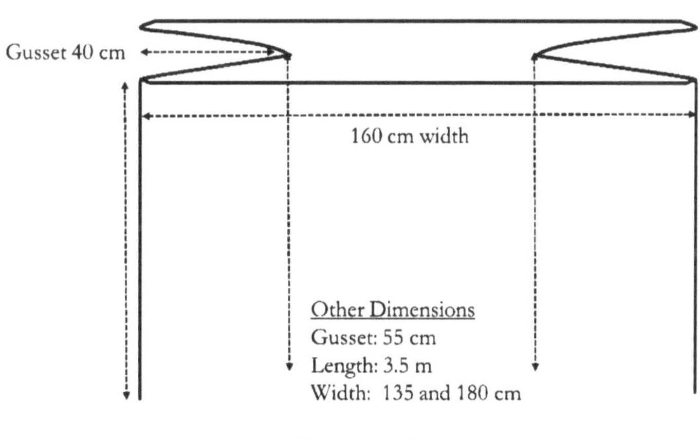

Figure 19.5

Gusseted silage bags

19.5 Bale silage stretch films

Coextruded higher α-olefin LLDPE stretch films are increasingly replacing bags to stretch wrap bale silage. The films need maximum strength in both MD and TD and are, therefore, preferably made by the blown film process. Coextrusion makes it possible to produce films with modified surface layers thus maximizing the differential cling properties enabling totally wrapped and sealed bales to be produced. Silage stretch films are 50 cm in width and supplied in 3000 m length rolls weighing 50 kg. The thickness is usually 25 μm. In some countries film width is increasing to 75 cm and as high as 100 cm. This makes it possible to reduce the number of layers required and allows for better overlapping with a reduced rate of oxygen permeation.

Silage film is a price-sensitive application which includes a significant proportion of re-granulated material. Blown film machines with 600 mm diameter dies with extruder outputs of around 500 kg/hr of resin are recommended to be competitive in this market. The PIB tackifier level is in the range of 8-9%. Thickness variation should be held at 10-15% from the target value on these large machines. Coextrusion is used to optimize cost and performance.

19.6 Soil fumigation films

These films contain fumigants which are laid on the soil to kill fungus, bacteria etc. The solarization film generates heat in the soil to destroy the "bugs". Fumigants such as methyl bromide are widely used as means of soil pest control. To improve the effectiveness of the fumigant and minimize atmospheric pollution, the soil is covered with a LDPE film. The gaseous fumigant is then introduced and the edges of the sheet are buried in the soil. LDPE films of 40-50 μm are normally used as covers. To reduce fumigant losses to the atmosphere, various multilayered films with improved barrier have been under evaluation in recent years. These include LDPE/HDPE coextruded films and structures based on polyolefins with barrier resins such as (co)polyamide and EVOH. **Table 19.5** shows some data at two temperatures comparing the permeation of methyl bromide (MeBr) through some typical films used in this application. The methyl bromide transmission rates are measured in g/m^2/hr. The EVOH is believed to be modified by a proprietary blending process. This data clearly shows the significant drop in transmission rates comparing the PE films and the barrier structures.

Table 19.5
Film permeation of methyl bromide

Film	Thickness (μm)	MeBr TR @ 20 °C ($g/m^2/hr$)	MeBr TR @ 60 °C ($g/m^2/hr$)
LDPE	35	22	75
LDPE/HDPE	40	2.15	-
PE/EVOH/PE	30	0.001115	0.03
PE/PA (copolymer)	35	0.122	0.5

The barrier films have been shown to reduce the level of fumigant needed for several crops. This reduction in fumigant such as methyl bromide is both environmentally friendly and economically attractive, thus off-setting the higher cost of the barrier films.

19.7 Solarization film

The usual procedure used to prepare soil for seeding is to disinfect with pesticides such as methyl bromide discussed above. However, in the advanced economies in particular, the heightened public awareness regarding the health risks caused by the use of pesticides in farming has led to the development of alternative methods for soil protection.

Solarization involves the use of transparent plastic film to cover the soil that has been well irrigated during the hotter months. When the ground heats up the combination of high humidity and high temperature leads to the destruction of harmful bodies which would attack the crop, e.g. insects, nematodes and fungus. Specially formulated LDPE films placed on the soil will allow temperatures of 145 °C to be reached some 15 cm below the surface. This intensive heat will destroy many of the harmful bodies in the ground without the need for undesirable chemical treatments.

19.8 Pond liners and geomembranes

The use of plastic film and sheet for water containment and soil support is growing rapidly in many regions of the world. These products are used in containment systems for water, hazardous waste and reclaimable materials. Butyl rubber sheet is also widely used but more costly than polyethylene films.

The application areas are:

- Pond liners used for water containment
- Geomembranes designed for soil support

Pond liners are utilized in specific end uses listed below:

- Waste water ponds
- Ditch linings
- Irrigation reservoirs
- Landfill floors
- Decorative ponds
- Evaporation ponds
- Farm ponds
- Shrimp and fish ponds

For the best chemical resistance the films are generally produced from carbon black-filled high density polyethylene (HDPE) in typical thicknesses of 1.5-3 mm. Tubular film lines capable of widths of up to 8 m are in use.

The liners are expected to last many years in exposed applications without any detrimental effect from sunlight. Liners should pass the low temperature brittleness test at -70 °C (-94 °F) according to ASTM D 746. In the case of landfills, in the US the Environmental Protection Agency mandates that all landfill liners must be 1.5 mm (60 mil) or thicker. 1 mm (40 mil) geomembranes are preferred for landfill closures.

Rolls are often 7 m wide and lengths range from 90 m up to 470 m depending on the material type and thickness. The gross weights can be approximately 7800 kg, and installation of pond liners requires trained workers and special equipment. The rolls are shipped by flat-bed trucks.

Films for pond liners and geomembranes must be resistant to chemical breakdown to ensure a safe barrier in use. The importance of containment, avoidance of seepage, stress cracking etc. will usually require strict quality control particularly during installation. UV light protection may also be required. The films are very often pigmented to inhibit plant growth.

Geomembranes films including liners are specified according to the following criteria:

- Relaxation time
- Stress-strain behaviour

- Chemical resistance
- Permeation
- Ageing
- Jointing

The selection of material for a liner application will depend upon the specific requirements and conditions encountered. For example in the multi-axial tension test, ASTM D 5617, HDPE liners rupture at 15% elongation and the LD/LLDPE liners rupture at 35% elongation.

HDPE liners are preferred for emergency containment of chemicals at high concentrations, such as fuels and acids. The HDPE liners provide a higher degree of chemical resistance than lower density polyethylenes.

Low density low modulus films will settle more readily when placed in a pond or landfill. However, if the PE density is too low chemical and permeation resistance are reduced. If the modulus or density is too high, impact and stress cracking resistance are poor and relaxation for settling may be inadequate. This leads to the conclusion that fractional melt index MDPE resins offer the best compromise solution in the most critical applications.

Weathering and ageing resistance is maximized by the combination of antioxidants, light stabilizers and carbon black. A maximum use temperature of 70 °C for HDPE and 60 °C for LD/LLDPE liners is usually specified.

A supplier of pond liners and geomembranes for containment must be able to produce wide rolls of sheet at varying thickness. Thickness will vary from 200 to 300 μm for pond liners. Some geomembranes may go up 0.5 mm in thickness. LDPE is still widely used because of the need for fractional melt index polymers with high melt strength to sustain large bubbles in the blown film process. LLDPE is used in lean blends to improve puncture resistance. Blend ratios rarely exceed 20% LLDPE. As discussed above fractional melt index MDPE is the best compromise product for these applications.

Coextrusions are used to improve scuff resistance where the ground is rocky or uneven. HDPE will be used as an outer layer in such constructions. HDPE will also improve water barrier. To increase flexibility and impact strength EVA, EMA or EnBA can be used.

To produce the large surface areas required, films are welded together for the required surface area. Three techniques are used:

- Hot wedge welding
- Extrusion welding
- Hot gas welding

The quality and strength of the join or weld is very important in order that the sheet meets specifications.

The types of PEs used in pond liners and geomembranes are listed in **Table 19.6**.

Table 19.6

Polymers for pond liners and geomembranes

PE type	MI (dg/min)	Density (g/cm^3)	Other factors	Usage
LDPE	0.3	0.918	-	Heavy duty grade
LDPE	0.3	0.923	-	Heavy duty grade
LDPE	0.7	0.922	-	General purpose
EMA	6.0	0.42	16-18% MA	Flexibility
LLDPE	1.0	0.918	C_4	Blended in LDPE
mLLDPE	1.0	0.923	C_6	Blended
MDPE	0.6	0.942	C_4	Geomembrane
MDPE	20 (21.6 kg)	0.940	Gas Phase	Geomembrane
MDPE	14 (21.6 kg)	0.933	Gas Phase	Geomembrane
MDPE	13 (21.6 kg)	0/938	Cr slurry	Geomembrane
HDPE	0.45	0.945	Broad MWD	Geomembrane
HDPE	0.05	0.950	Broad MWD	Coextruded

Some suppliers are offering ready-made black compounds. Light stabilizer (UV) formulated masterbatches are also available from several sources. Climatic conditions and the diversity of the applications must be taken into consideration when formulating film products for this complex and cost-sensitive market sector.

Chapter 20

Stretch wrap films

Stretch films are widely used to palletize and unitize loads as an alternative to shrink film or banding with metal or plastic straps or protective corrugated board. Four stretch wrap classifications are listed in **Table 20.1**.

Table 20.1

Stretch wrap classifications

Low stretch films	0 - 30% for manual/old machines
Conventional pallet	Up to 100% Low wrapping speed, standard
High performance	100-250% Specialty machine stretch
Ultra performance	250-300% High performance

The machine can revolve around the goods or vice versa, i.e. turning the pallet with the film dispensing unit in a fixed position.

There are five main market segments:

- Manual and conventional stretch film machines
- Pallet stretch: standard, specialty, ultra performance
- Silage bale stretch
- Bundle stretch
- Roll stretch/reels

A typical 1.5 tonne pallet is wrapped with 300-400 g of stretch film. This equates to 0.02% of the load illustrating the efficiency and environmental friendly aspects of this palletizing technique.

Silage stretch film is a rapidly growing segment of this film market. The film replaces bags for the bundling of grass feed (see Section 19.5).

The two basic stretch techniques for palletizing are shown in **Figures 20.1** and **20.2**. In the conventional process the rotating action of the load generates the force to stretch the film. The tension is controlled by the application of a brake on the film roll. In this technique all the force is exerted between the film roll and the corner of the load. High stretch levels may crush the load or move it on the turntable. Practical stretch levels are restricted to 20-55% which is relatively inefficient since most films can achieve much higher levels of elongation.

In the pre-stretch method (**Figure 20.2**) the stretching is isolated from the corner of the load. The film is passed between two rollers rotating at different speeds. The second roller has a higher velocity, which stretches the film in the gap between the two. This method can easily achieve stretch levels of 250%. The higher degree of orientation obtained reinforces the film and reduces its thickness significantly. Yield is increased per load. This method also allows the operator to independently control the stretch tension thus maintaining consistency throughout a shift.

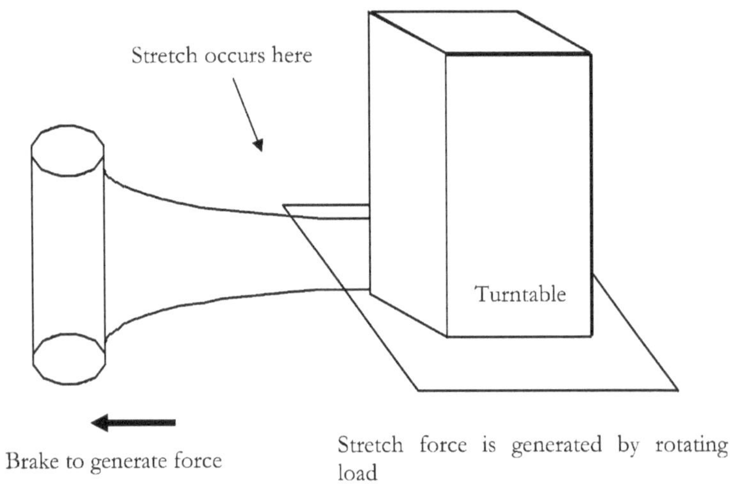

Figure 20.1

Conventional stretch process

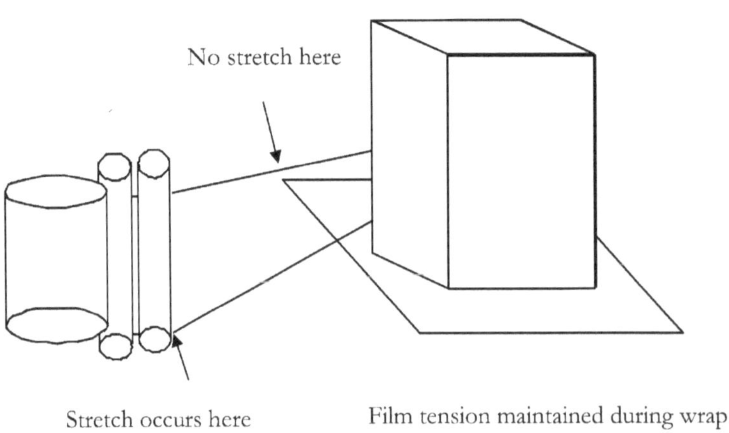

Figure 20.2

Pre-stretch method

A more recent development is the stretch hood. This is a polyethylene film tube, usually a 3-layer coextrusion, sealed at one end and stretched over the palletized load. The film is cut to the required length and gathered on four grips, which stretch the film in the horizontal direction. The film is stretched until the dimensions are slightly larger than the load. The film is pulled over the pallet unrolling it as it moves over the load. By controlling the unrolling rate the film is stretched in the vertical direction. This holds the load on the pallet. At the bottom the grips release the film which then wraps under the pallet bottom. The film must have sufficient slip to be able to slide over the load without wrinkling. Film thickness is 110-120 μm. In the coextrusion the core layer is made up of blends of EVA and POE. Outer skins of LLDPE or mLLDPE are recommended. This technique is claimed to offer a faster rate of palletization over conventional stretch wrap.

LLDPE has reached a penetration rate of ~90% of the stretch film market. Coextrusion is used to optimize properties at the lowest cost. The balance is made up largely of EVA and some PVC for soft/fragile goods. Economics are influenced principally by the film weight per load. The performance criteria that determine the quality of stretch films are described below:

1. Factors influencing the breaks per roll
    - Film gauge
    - Thickness uniformity
    - Splits/nicks in the film
    - Gels and other defects
    - TD tear strength
    - MD ultimate elongation
    - MD ultimate tensile strength
2. Factors influencing load retention
    - Cling to non-cling surfaces
    - Cling-cling surfaces
    - Cling retention with time
    - Puncture resistance
    - Creep resistance
    - Force to load/stretch

The polymer related properties that are perceived as most important to film processors are:

- Puncture resistance
- Tensile strength
- Tear resistance
- Elongation
- Clarity/low haze

To minimize cost the material choice should be based on downgauging and high elongation to achieve maximum coverage. Most films are coextruded, sometimes containing up to seven or more layers. The formation of coextruded microlayers is claimed to result in superior properties which enable downgauging with optimum choice of materials and blends.

20.1 Method of production

Both blown and cast films are used for stretch wrap packaging applications. The breakdown is 80% cast film and 20% by tubular film extrusion. Tackifier resin, usually PIB (polyisobutylene), is added during the film extrusion process.

The cast film process is well suited for high volume markets particularly where coextrusion is required. Coextruded films from 3 to 30 microlayers are produced. Five layers from four extruders are now common. Some machines are capable of operating at up to 700 mpm line speeds. Thicknesses can range from 8-76 μm for machine and hand wrap films.

Films produced by the casting process have superior optical properties, lower neck-in at stretching and higher TD tear resistance values compared to blown film. Chill roll contact produces a very smooth film surface helping to generate the cling effect. However, the rapidly growing silage market, which requires high puncture resistance, is better suited to the blown film process, which yields tougher films. Low BURs should be used. This improves machine direction stretching and enhances tack.

20.2 Coextruded stretch film structures

Some established stretch film structures are shown in **Figure 20.3**. Coextruded films are usually configured to include a thick low cost and high strength core layer with outer skins of more expensive tacky polymers. One side cling effect can be designed in the structure with PP homopolymer as a skin layer in a coextruded structure.

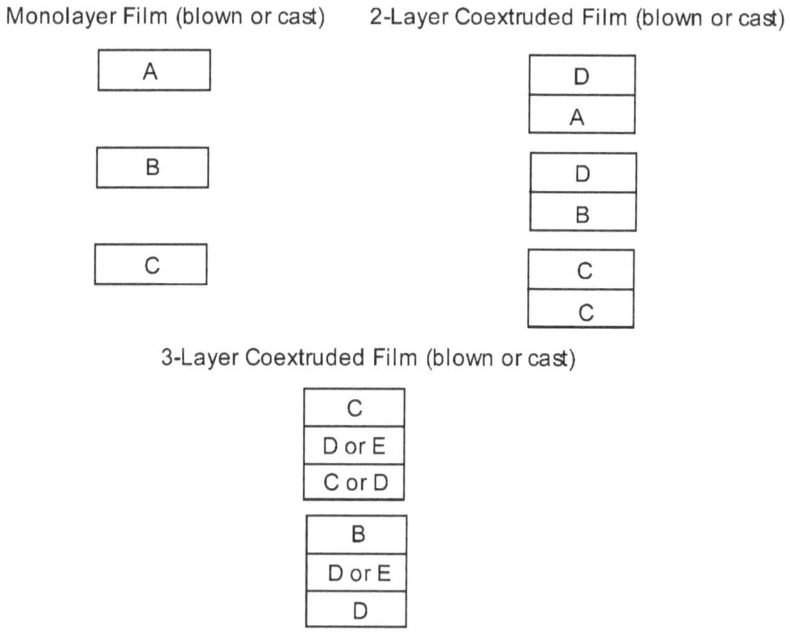

Figure 20.3

Stretch film configurations

A = EVA (12% VA)
B = LLDPE with low molecular weight tackifier
C = LLDPE with polymeric tackifier
D = LLDPE
E = LLDPE + scrap (post production)

Figure 20.4 and **Table 20.2** describe typical 2-layer cast film structures based on LLDPE.

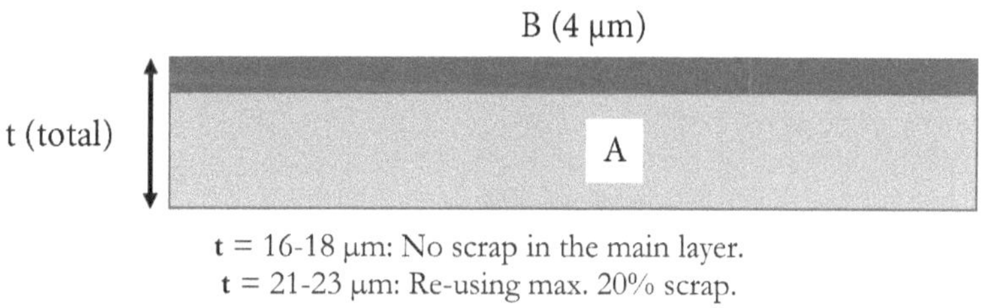

Figure 20.4

Coextruded LLDPE in 2-layer stretch films

Table 20.2

Optimum polymer combinations

Film	LLDPE	MI (dg/min)	Density (g/cm^3)	Additive
Layer A Layer B	C_4 C_4	2.5 1.0	0.919 0.919	tackifier -----
Layer A Layer B	C_8 C_8	2.3 2.5	0.917 0.935	----- tackifier
Layer A Layer B	C_4 PP cop.	2.5	0.919	tackifier -----

The following extruder sizes and outputs are typical:

Extruder A 70 × 25 D

Extruder B 120 × 30 D

Die width 2100 mm

Line Speed 100-135 mpm

Thickness 16-23 µm

Output 300-350 kg/hr

The alternative 3-layer structure in **Figure 20.5** will typically be made according to **Table 20.3**. Extruders #1 and #3 could be downsized to 2½" extruders (6.35 cm). Extruder #2 needs a long screw (32 D) in order to ensure complete homogenization of the 30% scrap material.

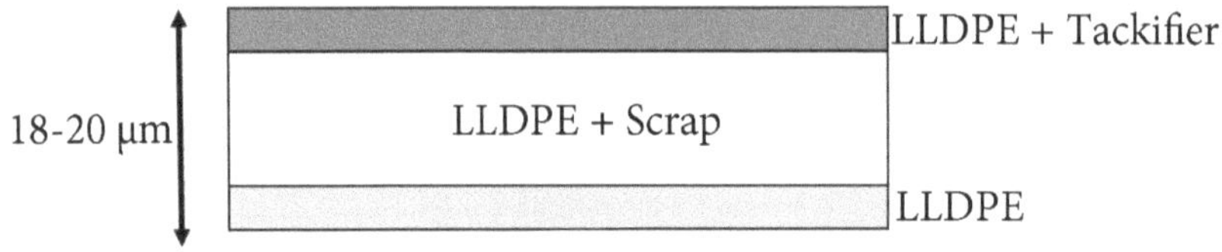

Figure 20.5

Coextruded cast film structure

A formulation for a high performance 300% pre-stretch, 2-layer cast coextruded film based on mLLDPE is suggested below.

Layer 1: mLLDPE (0.918 g/cm^3/3.4 dg/min. MI): 10%.

Layer 2: As above blended with 10% EVA or mPE: 90%.

Recommended thickness: 17-23 µm.

Table 20.3

Coextruded cast film formulation for stretch films

Extruders	**Polymer**	**MI (dg/min)**	**Density (g/cm³)**	**Max output (kg/hr)**	**Remarks**
#1 3½ " (8.9 cm) 28 D	C_8 LLDPE, Tackifier	2.6	0.920	170	Excellent elongation, puncture, low tear propagation
#2 5" (12.7 cm) 32 D	C_4 LLDPE 30% scrap	1.0	0.918	480	Strength, cost optimized
#3 3½" (8.9 cm) 28 D	C_8 LLDPE	1.0	0.920	180	Optimum toughness

The stretch film industry has recognized that increasing the number of layers will enhance stretchability, tear and puncture resistance. This is true even if the same PE grade is used. The layer multiplier technique available with feed block coextrusion creates microlayers, which can save on raw material costs. Breakages caused by pinholes, gels and other defects are eliminated as the layers are encapsulated and rendered harmless. Thus many of the basic examples above have been superseded with more sophisticated proprietary recipes, with infinite possible combinations of materials and layers.

20.3 Properties of stretch films

Table 20.4 compares commercial stretch films extruded from three LLDPEs. The films were 20 μm in thickness and 0.5 m wide.

Table 20.4

Properties of LLDPE stretch films

Property	Units	Butene	Hexene	Octene
TD Tear	g/μm	560	920	1000
MD Tensile strength	MPa	21	27	24
MD Elongation	%	480	570	460
Puncture @ pre-stretch of 50%	kg	1.0	1.0	1.0
Elongation @ pre-stretch of:	%			
50%		200	200	200
70%		350	400	350
Max. pre-stretch before break	%	75	80	80

Table 20.5 compares the film properties of a gas phase hexene LLDPE with solution octene LLDPE. This data shows the close match between the two LLDPEs. The hexene grade has a lower density.

Table 20.5

Comparison of gas and solution phase LLDPE

Property	Units	Gas phase 1-hexene	Solution 1-octene
Melt index	dg/min	1.0	1.0
Density	g/cm³	0.918	0.920
Secant modulus MD/TD	MPa	204/233	200/214
Tensile strength MD/TD	MPa	40/34	45/35
Dart impact strength	g/μm	7.1	6.7
Tensile impact MD/TD	MJ/m³	161/147	150/125
Elmendorf tear MD/TD	g/μm	14/27	15/28

20.4 Household cling films

Plasticized vinyl cling films have been in commercial use for around 50 years in Europe and N. America. These films are used for wrapping fresh meat and poultry in supermarkets, for food storage in catering outlets and for food protection in the home. The market was dominated by plasticized PVC and PVDC (Saran Wrap). Polyethylene films have penetrated this market initially to counter concerns over plasticizer migration. Over time these lower cost films have established themselves in this market, but they are generally accepted to be inferior in their cling and stretch properties to the vinyl films.

Table 20.6 shows the average properties of PE household cling wrap films typically found in Europe. These are extruded principally from LLDPE and a cast film process is often used. Cling additive may be added, typically 0.5-1.0% liquid PIB. Film thickness has been reduced over time to around 11 μm.

Table 20.6

Properties of LLDPE kitchen wrap films

Properties	Unit	Typical spec.
Average thickness	μm	11
Tolerance	±%	9-12
Tensile strength MD/TD	N/mm²	32/23
Elongation MD/TD	%	160/550

The household wrap market is less demanding than point of sale cling wrap used in retail outlets, where very high transparency, quick seal and spring-back (recovery) are important requirements. Soft vinyl films are still the best products both for retail and household markets. The PVC films are also easier to tear on the serrated edges of the cardboard container compared to LLDPE films.

20.5 Retail cling films

PVC films are still preferred for the more demanding retail applications and especially with automatic machines because of their excellent machineability, snapback and cling. Key film characteristics are:

- Clarity
- Low extractability (< 10 mg/dm^2)
- Tear strength
- Cling
- Finger indent recovery (snapback)
- Machine handling
- Antifog (for fresh meat and produce)

20.6 Production methods for cling films

EVA film (± 10% VA) was the first substitute for PVC, but tear strength, finger indent recovery and cling were poor. This was improved by using higher VA content grades plus tackifier. However regulations prevent the use of films containing more than 10% VA in direct contact with fatty food products, which includes fresh meat in certain countries.

An alternate product is a coextruded LL/LDPE (80%) and EVA (20%) film. A further refinement is coextrusion of EVA with skin layers of mPE (density <0.91 dg/min.). The metallocene resin has very low extractables, high elongation and low haze. The metallocene PE films will cling, but not block or stick. Films with a better balance of MD/TD properties are claimed with mPE.

To achieve the highest clarity, the cast film process is utilized. Cast LLDPE with a monomeric tackifier is another technique. GMO (glycerol mono-oleate) has traditionally been added to PE films to enhance cling properties and to act as an antifogging agent. In order to preserve the appearance of the package it is important to prevent moisture from coalescing into tiny droplets creating a hazy effect on the film's internal surface. Another cling additive used is sorbitan mono-oleate. Food wrap cling film resin formulations based on EVA and atactic PP (APP) are also available. The APP is specially compounded and produced in pellet form for blending with the host resin. The availability and quality of APP can be variable.

Due to the specialist nature of stretch films, winder design is critical for high quality rolls. The most important factor in achieving good roll quality is very accurate tension control within the line, and particularly so within the winder assembly. It is usual to use an S-wrap nip rather than contact nip rollers in such applications. For web separation a driven separation nip is required to reduce the risk of tension variation and "plucking". To achieve rapid response and enhanced operating range, drivers should be AC flux vector and tension should be controlled by load cell and not dancer rolls.

If blown film is used the bubble cage and collapsing section need to be of a low friction design, preferably non-contact. With stretch films, cooling in the primary nip assembly is recommended to reduce the blocking tendency.

The preferred winding technology is centre-winding/gap winding and some form of automated roll handling is preferred because of the short time between roll changeovers when producing the standard 300 mm or smaller diameter rolls. To reduce vibration and flexing of the winder caused by high speed winding of sticky films (>150 m/min.) the winder must be of substantial construction.

Slitting in-line can only be successfully accomplished by sharp blades arranged to minimize the chance of the film grabbing the blade. Star type blade holders and twin blade holders have been used successfully for automatic blade changes on these films.

20.7 Tackifier and cling additives

The most common tackifier used in stretch and cling film is PIB (polyisobutene). The migration of the PIB to the surface of the film is referred to as blooming, and the rate at which migration to the surface occurs is referred to as the bloom rate. The amount of time required for this migration to occur to achieve optimum cling properties is known as bloom time. Problems encountered with PIB blended with polyethylenes are:

- Handling due to high viscosity
- Inconsistent bloom rates
- Long bloom times

- Smoking during processing
- High noise levels during wrapping
- Condensation on air rings
- PIB loss during processing

PIB is a difficult polymer to blend with PE and the usual method of addition is by injecting molten low molecular weight PIB in the melt stream just before the mixing section of the screw. In blown film 6-9% is added depending on the application. The concentration in cast film can be reduced to 0.5-2% PIB for some applications. There are a number of proprietary mixing systems that can be incorporated in the extruder.

An alternate method is adding the PIB from masterbatches. This process eliminates the need for specially designed extruders with liquid injection sections plus heated storage tanks for the PIB. These concentrates (15-20% PIB) are produced by proprietary twin screw extrusion processes, which disperse high concentrations of low MW (liquid) PIB in PE resins (LL/LDPE). Some compounders have succeeded in incorporating up to 50-60% PIB in the carrier resin. Obtaining free flowing pellets at these high concentrations is clearly challenging and process details are kept secret.

In order to optimize performance, blends of both low and medium to high molecular weight PIB are dispersed in the resin. The key aim is to maximize speed of migration to the films surface and to develop permanent cling. Permanence is provided by the high MW PIB component. However, excess high MW PIB can cause high noise levels when unwinding in stretch applications. It is claimed that the use of atactic PP (APP) can reduce noise. Formulation expertise has clearly increased in sophistication over the years.

The PIB polymers used for cling applications are listed in **Table 20.7**.

Table 20.7

PIB grades used for cling applications

Molecular weight	**Density (g/cm^3)**	**Viscosity (mm^2s^{-1} @ 100 °C)**
950	0.894	225
1300	0.902	635
2300	0.911	2565
2600	0.914	4250

The final PIB concentration, irrespective of method of addition is specified by the application requirements as shown in **Table 20.8**.

Table 20.8

PIB addition levels by application

Application	Blown film	Cast film
Silage wrap	6-8%	4-6%
Pallet stretch	3-6%	2-4%
Domestic cling film	2-4%	1-2%

Alternatively EVA with VA contents between 12 and 18%, and in some cases atactic polypropylene (APP) from masterbatches has also been evaluated as tackifier. The APP masterbatches are claimed to avoid the risk of screw slippage which may occur with liquid PIB. Metallocene PE skin layers in coextrusions have also been considered as an alternative to replace PIB-tackified LLDPE. Cost is the usual deciding factor.

Other options for the film converter are low molecular weight non-polymeric agents such as glycerol mono-oleate. These migrate to the surface after a certain storage time. The disadvantages of using non-polymeric cling additives are:

- Unpredictable time and temperature influence
- Migration to both film surfaces in coextruded films

Chapter 21

Shrink films

Heat-shrinkable films are used to provide a protective wrap to unitize individual items, to protect food products and to palletize heavy shipments for transport and distribution.

Shrink packaging offers the following advantages over other packaging technologies:

- Economically unitizes multiple items for shipment
- Protects the product from dust, moisture and damage
- Provides a contour fit allowing all shapes to be wrapped attractively
- Discourages pilfering and shop lifting
- Adds gloss/sparkle enhancing product appearance
- Cuts labour costs
- Reduces labelling costs
- Increases production speed

In most packaging applications dimensional stability of the film is a key asset, however, in this market segment controlled shrinkage of the film at elevated temperatures is necessary. Over the years a broad range of heat-shrinkable plastic films have been developed to meet the requirements of many end-use applications. The most ubiquitous are those based on polyethylene films. These films are produced by two basic processes:

- Tubular/blown film process
- Tubular double bubble film process

21.1 Production of tubular LDPE shrink films

LDPE shrink films are made in a single step by the blown film process. Shrinkage is controlled by the blow-up ratio and draw rate between the die gap and the final film thickness. These films are used in industrial palletization, consumer unitization and bundling applications. Thickness ranges from 30-180 μm depending on the application. Shrinkage for this type of PE film is in the range of 15-20%, which is acceptable for many retail and industrial applications, but may be insufficient for some applications requiring a very tight attractive package for point of sale display.

Blown LDPE film is the lowest cost shrink film with an attractive combination of properties and is, therefore, the most widely used in industry and consumer packaging. Shrink temperature for LDPE is in the range of 110-120 °C. In practice the oven air temperature in the tunnel can be as high as 140 °C.

Table 21.1 subjectively compares the performance of commonly used shrink films for consumer packaging. If higher shrink tension is specified, 5-10% MMW-HDPE may be added to LDPE.

The conventional blown film PE process is used to produce films over a thickness range from 30 up to 210 μm. Small dies are utilized if high blow-up ratios (>2.5:1) are specified. Films produced by this method will not be as strong as the biaxially oriented films using a double bubble process.

Table 21.1
Comparison of shrink film performance

Film type	Advantage	Possible problems
LDPE/LLDPE lean blends	Strong seals	Narrow shrink temperature range
	Low temp. shrink	Low stiffness
	Medium shrink force	Moderate opticals
	Broad application	Sealing wire contamination
	Lowest cost	
Coextruded films	Good opticals	Cost/property optimization
	Good machineability	Superior film property balance
	Low shrink temp.	Wide choice of recipes
	Optimum seals	

LDPE resins of MI 0.3 to 2.0 dg/min. are used. Pure LLDPE is not recommended in this application because of its very low shrinkage in the TD even at high blow-up ratios. This is caused by the necessity to extrude LLDPE with dies using wide die gaps in order to avoid sharkskin and excessive pressure build-up in the machine. Because of its linear structure LLDPE is more prone to creep under tension and this may cause the wrap to loosen over time. However, new metallocene grades are now available which can be blended with LDPE or used alone, which enhance shrink film properties. These have fractional melt indexes of 0.3 dg/min and densities from 0.920 to 0.935 g/cm^3.

The widening of the die gap to avoid sharkskin with LLDPE has the following ramification for film orientation. In **Table 21.2** the effect of three die gaps on film orientation for a 30 μm film processed at 2:1 blow-up ratio is illustrated. A 300 mm diameter blown film die is tested with a mass flow rate of 200 kg/hr at three die gaps. This calculation assumes a melt density of 760 kg/m^3 from which the velocity of the melt at the die is derived.

Table 21.2

Effect of die gap on film orientation

Die gap (mm)	0.50	1.00	2.50
BUR	2.00	2.00	2.00
Velocity @ die (mpm)	9.31	4.65	1.86
Velocity of film (mpm)	65.00	65.00	65.00
Total stretch ratio	6.98	13.98	34.95
TD Stretch ratio (BUR)	2.00	2.00	2.00
MD Stretch ratio	3.49	6.99	17.48

The MD stretch ratio increases from 3.49 to 17.48 or roughly times 5 as the die gap is widened from 0.5 to 2.5 mm. Processors will normally use the blow-up ratio to determine shrinkage. These calculations show that the die gap setting will also impact the outcome.

In addition, because of its low melt strength, maintaining bubble stability with LLDPE becomes increasingly difficult at high blow-up ratios. The exception is pallet pass through and MD oriented collating film, where LLDPE has better elongation and balanced shrink is not required.

LLDPE can be blended with LDPE to increase shrink tension, resistance to burn-through (which can result in tearing) and seal strength. Addition of HDPE will increase shrink tension for heavy loads. Shrinkage can vary from 50-70% in MD to 5-15% in TD at normal (2:1) blow-up ratios. More balanced shrinkage in MD and TD is obtained by increasing blow-up ratio up to 4:1 when required. EVA is used for certain applications to achieve the following:

- Tack
- Low temperature sealing
- Low temperature shrinking
- Low shrink tension

EVA shrink films are useful for collating crushable or fragile items that may be damaged with excessive shrink force. This is a segment of the market once dominated by plasticized PVC. LDPE is the dominant material used in shrink packaging and heavy duty palletization.

The BUR for shrink films fall into two categories.

1. BUR ~1.5:1. This yields films with 60-80% shrinkage in MD and approx. 10% in TD.
2. BUR ~4:1. Yields reasonably balanced shrinkage of 30-40% in MD and TD.

The films with high MD shrinkage are used in collation applications such as cans where one way shrink is required. Typical recommended BURs for different market segments of thin shrink films are as follows.

Segment	**BUR**
Rolls, e.g. carpets, wallpaper	1:1.5
Bottles/cans collation	1:1.5
Cucumbers	1:1.5
Cartons, boxes, stationery	1:3-4

Coextrusion is increasingly used to optimize performance and cost. Some processors claim superior properties with 3-layer coextrusions and others use 5 layers. A variety of blends with LD/LLDPE and metallocenes are increasingly used. A 40 µm 3-layer coextruded film will typically have the following specification:

- Skin layers: blend of 85% mLLDPE + 15% LDPE (0.5 MI)
- Middle layer: blend of HDPE (0.7 MI/0.960 D) + 35% LLDPE (C_4)

The HDPE adds stiffness and eases the processability of the LLDPE. The layer ratios will vary with the application requirements.

21.2 Double bubble process

This process is used to produce low gauge high clarity films for point of sale wrapping and specialized food packaging. Strong, glossy, thin films are produced by this technique. Shrinkage is balanced and normally in the 50% range. The film produces very tight attractive and glossy packs. These films are used to pack high value added products including sausage casings. The double bubble process makes it possible to induce orientation in linear polyolefins at precisely controlled conditions in relation to their first order transition temperatures. Low gauge, high strength and transparent films are produced with LLDPE and polypropylene (PP/PPRC) by this technique.

The double bubble process can accurately control the stretch temperature and ratio along the two principle axis, which results in films with excellent physical and optical properties. Perfectly balanced MD and TD shrinkage is possible. Very strong and glossy films down to 12 μm in thickness are produced by this technique. Development work is ongoing to further downgauge to 8 μm. A typical stretch ratio is 6.6 × 6.6, starting from 1.2 mm thick film in the primary bubble.

The double bubble process to produce biaxially oriented films is shown in **Figure 21.1**. Some plant designs have the extruder feeding downwards with both bubbles aligned vertically.

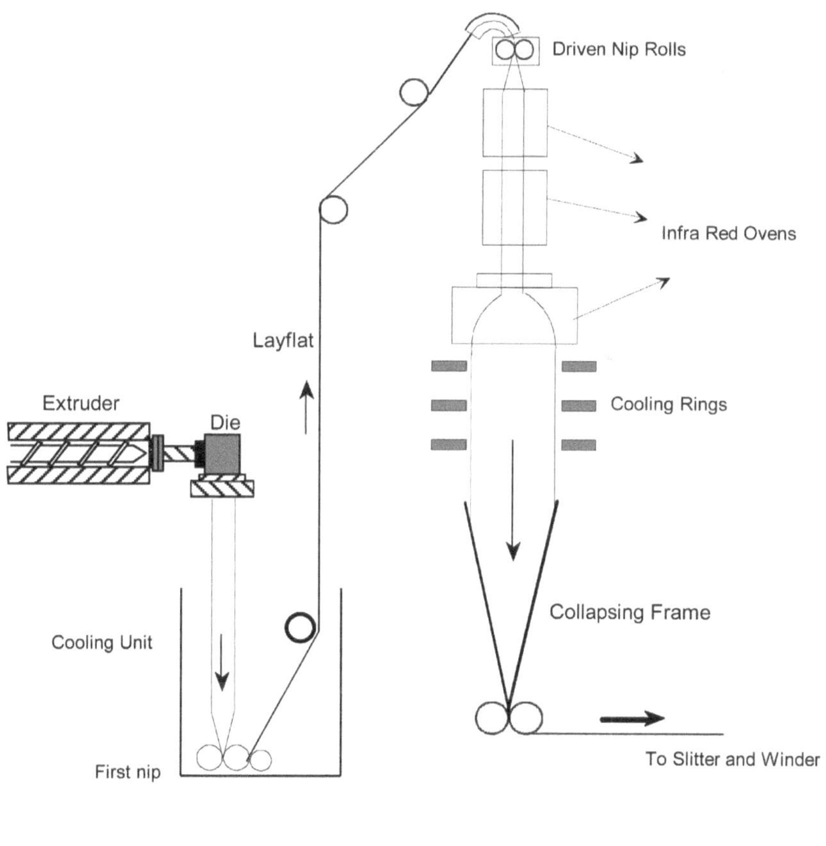

Figure 21.1

Double bubble process for biaxially oriented films

In the first step, the polymer is extruded into a tube and drawn into the cooling unit. External cooling is carried out in a water cascade or bath and the internal film surface may be cooled with proprietary devices such as chilled mandrels. The cooling temperature ranges from 15-30 °C. The temperature variation around the tube circumference should be no more than 1 °C.

Film thickness is in the range of 150-1500 μm and the degree of orientation is minimal at this stage. The cooled thick layflat tube is then pulled through a second nip and reheated in infra-red or air heated chambers. By injecting air the layflat is re-inflated at the appropriate temperature and bi-oriented. It is claimed that the bubble can be held in continuous operation for one month. With LLDPE the temperature of the film in the orientation oven is 100-115 °C. If polypropylene (PP) is used the air temperature is raised to around 170 °C. The MD orientation is controlled by the speed differential between the second set of nip rolls and the haul-off equipment. A typical speed differential is 5 mpm in the oven and 26 mpm at the haul-off nip. The blow-up ratio determines the degree of TD orientation. The air cooled bubble is then collapsed into a layflat as shown and transported to the winding equipment. The haul-off unit is normally rotated or oscillated to re-distribute any thickness variations across the film roll. The rolls of film are usually conditioned for a few days prior to converting operations such as printing and slitting.

A feature of this process is that the second bubble does not come into contact with any machine parts, which eliminates the formation of scratches or other surface defects on the high gloss film. This enables the processor to use lower melting polymers as outer skin layers if required. The final film thickness can range from 12 to 38 μm with a claimed variation of ±1 μm. Typical outputs will be as follows:

- Film width of 1.8 m @ 145 kg/hr
- Film width of 2.4 m @ 220 kg/hr

If a heat set (annealed) film is required, the single-layer film is passed through a heated tenter frame and held under stress during the annealing process. On some machines a third bubble is formed for annealing. A more dimensionally stable film is produced by this additional treatment. BOPP films are usually heat set by an annealing process irrespective of the orientation technique used. The throughput of the annealing equipment is usually much higher than the bubble orientation process, which makes it possible to feed one unit from two or more extrusion lines thus spreading the extra investment costs. Heat setting of BO-LLDPE films is, however, rarely done.

Film properties can be modified by coextrusion. These can combine advanced polyolefins based on metallocene technology and other processes. Copolymers based on propylene with ethylene and other olefins are used where higher melting temperatures are specified.

A typical coextruded film is formulated as follows. These will shrink at 120 °C.

PP terpolymer (15%)
LLDPE (70%)
PP terpolymer (15%)

The advanced PP terpolymers have heat-seal initiation temperatures of 110 °C and can provide high transparency films. They also have higher stiffness than LLDPE but are costly.

21.3 Properties of BO-LLDPE films

The deformation mechanism of polymers is based on the following three phenomena:

1. Elastic deformation, which is recoverable when the stress is removed.
2. Viscous deformation caused by molecular slippage, which is non-recoverable.
3. Molecular alignment caused by uncoiling, which is frozen-in on cooling and recovered on heating.

For high quality shrink packaging applications where fast, balanced and predictable shrinkage is required, the film needs to be oriented at condition 3. By controlling temperature in the second bubble recoverable orientation is induced in the film.

Orientation is easier with linear polymers and as a result, this type of technology has been further advanced with the availability of LLDPE. This was first exploited by DuPont with their range of Clysar® shrink films (Clysar is now an independent company). High α-olefin LLDPEs allow higher stretch ratios than butene grades and will yield stronger films and more balanced orientation. Forming and stabilizing the second bubble is not easy. There are patents that claim that the melt should be strain hardened to stabilize the second bubble[1]. In the Cryovac® process (www.cryovac.com) the tube from the first bubble is lightly crosslinked by electron irradiation, which increases melt strength. However, as mentioned

[1] Baird, W.G., Lindstrom, C.A., Besse, A.L., d'Entremont, D.J., Method of producing film having improved shrink energy, US Patent 3,022,543, 1962, and Benning, C.J., Gregorian, R., Werber, F.X., Process for forming cross-linked polyethylene film, US Patent 3,201,503, 1965

earlier LLDPE has been successfully commercialized into excellent shrink films by this technique. The development of advanced metallocene catalysts has extended the polymer options for tailoring film properties to a variety of applications.

Table 21.3 compares the properties of a conventional LLDPE single step blown film to a film produced by the double bubble process from the same octene LLDPE. All the properties except the Elmendorf tear propagation are very significantly improved by the double bubble technique. The improvements in opticals are dramatic.

Table 21.3

Comparative properties of BO-LLDPE film

Property	Unit	ASTM	LLDPE	BO-LLDPE
Melt index	dg/min	1238	0.9	0.9
Density	g/cm^3	1505	0.9230	0.9230
Orientation ratio	MD/TD		-	6/6
Thickness	µm		15	16
Gloss 45°	%	2457	58	89.2
Haze	%	1003A	9.6	1.22
Dart drop	kJ/m	1709	29	111
Elmendorf tear MD/TD	kN/m	19922	88/373	8.7/7.7
Sec. Modulus	MPa	882B	232/280	535/528
Tensile @ break MD/TD	MPa	882B	47/34	99.2/112.5
Elongation @ break	%	882B	528/780	100/100
Puncture resistance	N/mm		1656	7745

A series of high quality biaxially oriented shrink films are compared in **Table 21.4**. These include PVC, blends of LL/LDPE and irradiated LLDPE of the Cryovac type. In the Cryovac process an extruded thick tubular film is irradiated in-line with the extruder by electron bombardment usually at a dose below 10 Mrads. The tube is reheated and biaxially oriented in a second bubble. The effect of irradiation is to induce molecular crosslinking, which increases tensile strength and shrink tension and resists burn-through. The presence of crosslinking accelerates the orientation process and the rate of shrinkage in use.

Table 21.4

Comparative properties of commercial shrink films

Property	Unit	BO-LLDPE	OPVC	BOPP (PPRC)	XL BO-LLDPE	BO-LL/LDPE
Melt index	dg/min	0.9		3.9	1.0	1.0
Density	g/cm³	0.923	1.40	0.8997	0.9240	0.922
Orientation ratio	MD/TD	6/6		6/6		
Thickness	µm	16	18	15	15	15
Gloss 45°	%	89.2	81	83.3	89.0	92
Haze	%	1.22	4.4	2.5	2.07	1.4
Dart drop	KJ/m	111	119	194	184	150
Elmendorf MD/TD	KN/m	8.7/7.7	8/21	1.36/1.79	11.1/11	12.8/11.5
Secant modulus MD/TD	MPa	535/528	855/792	1656/1371	340/300	270/250
Tensile @ break MD/TD	MPa	99/112	56.5/49	81/80	83/87	83/93
Elongation @ break	%	100/100	73/155	10/20	70/80	140/130
Puncture resistance	N/mm	7745	3880	7140	3490	3706

XL: Crosslinked by irradiation. Commercial product.
BO-LL/LDPE blend: Commercial film. PPRC: PP random copolymer.

21.4 Markets for BO-polyolefins

Shrink packaging is used in the consumer goods market for the following display items:

- Wallpaper
- Gift paper
- Stationery and graphic products
- CDs and DVDs
- Software products and documentation

- Cosmetics and pharmaceuticals
- Cleaning products
- Frozen food
- Box packing, toys, puzzles, etc.
- Promotional packaging
- Food: cheese, vegetables, bread, etc.

Examples of shrink films produced by the double bubble process are shown in **Table 21.5**.

Table 21.5

Biaxially oriented films

3-layer	PPRC/PP/PPRC
3-layer	PP/LLDPE/PP
3-layer	LLDPE/PP/LLDPE
3-layer	PP terp./LLDPE/PP terp.
Mono	LLDPE

The fine shrink market is subdivided into three categories. The high shrink films offer the strongest competition to PVC films.

Low shrinkage	8-15%
Medium shrinkage	20-30%
High shrinkage	40-60%

Low shrink films are used for bread and other bakery goods. Bread wrapping is a large market in France and Italy. These films are also supplied in Germany and UK, but are not usually shrunk. Prices are at the bottom end of the range.

Medium shrink films are used for wrapping a variety of products including bread, loose toys, magazines, etc. Prices are in the middle of the range quoted.

High shrink films are used for wrapping stationery, chocolate, biscuits, fruit trays, toys, games, picture frames, candles, wallpaper rolls. These are often LLDPE-based and include blends and coextrusion with EVA, PPRC, PO terpolymers, etc. These products come closest to matching PVC shrink films, where low shrink temperature and low shrink tension are specified. These films are at the high end of the price range.

Chapter 22

Heavy duty bags

Heavy duty bags are produced in various forms:

- Valve bags
- Open mouth bags
- Fill/seal gusseted bags
- Glued square bottom bags

They are typically 300-610 mm wide and 100-200 μm thick.

22.1 Heavy duty bag design

Heavy duty PE sacks are utilized to contain and ship such items as fertilizers, mulch, potting soil, rock salt, powdered chemicals and plastic resins. The bag has to protect the product from any exterior damaging influence and has to be resistant to any aggressive reaction from the packed product itself. For consumer applications the sack is generally designed to hold weights from 5 to 10 kg although for some consumer and commercial uses the sacks are specified to contain up to 25 kg of product. The bags are pre-fabricated tube cuts with a bottom seal designed to be filled in fill and seal equipment. The alternative is a tube roll supplied to a form, fill and seal machine. The valve bag is a square bottomed prefabricated bag, which is also widely used for free-flowing products such as plastic pellets. Glues are used to form the square bottom and to attach the self-closing valve system. The valve bags will square on filling exhibiting greater stability on the shipping pallet. Key film properties include the following:

- High tensile strength
- High puncture resistance
- High work absorption power during dynamic stress
- High seal strength
- Low slip level
- Heat stability during filling with hot material

- Moderate stiffness
- Scuff resistance

In addition to the above, the sack should be moistureproof and often allow for short term outdoor storage. Protection from moisture and see-through visibility are important advantages that the PE sack offers over multi-wall kraft paper sacks.

Heavy duty bags are typically produced on blown film lines with small diameter dies. These lines have to be run at the highest output, and will usually require an IBC system with an efficient cooling ring. Fractional melt index (0.2-0.8 dg/min) PE grades are used.

Extruders with either smooth bore or grooved feed extruders can be used. Bubble cooling is usually the limiting factor on output and not the extruder. Techniques such as water cascades on bubbles extruded downwards have been developed over the years but these have gained limited acceptance. Air cooling is the principal method used to control the process. Most of the sacks are printed and colour masterbatches are added to produce an opaque background when required. Screws with good mixing capability must be used to ensure the most efficient pigment dispersion.

The design of the nip from the collapsing frame is critical for heavy duty sacks. To prevent blocking additional cooling within the nip assembly is recommended. More importantly the design of the nip itself is critical if the edge folds are not to become weak spots. In a standard primary nip, if the crease formed at the nip is too sharp splittiness may occur. The correct pressure must be applied at the nip to avoid excessive creasing.

Fractional melt index LDPE film has been proven to be the best film to comply with the tough demands made on the filled bag containing up to 25 kg of product. Melt index ranges from 0.2 to 0.8 dg/min. and density ranges from 0.918 to 0.921 g/cm^3 have been the standard. The classic heavy duty bag is made from 220 μm thick blown film using 0.3 melt index LDPE although films are now increasingly made from LD/LLDPE blends and thickness has been reduced to 180 μm. The higher the strength demanded, the more LLDPE is utilized. Downgauging to 120 μm is claimed with LLDPE rich blends. HAO-LLDPEs are preferred for maximum strength. Downgauging has also been claimed for 3-layer coextruded films using various resin combinations including recyclates in the different layers. Coextrusion also offers the advantage of varying the surface characteristics of the inside and outside of the sack to optimize stacking performance.

The use of mLLDPE presents no additional problems for equipment selection, with the exception of cutting the film at the end of the roll or bag if made in-line.

Because the metallocene polymers are tougher, any cutting device needs to have sufficient power to cut through the thick film and this is particularly important if the sacks are gusseted. A flying knife system is used successfully for most situations. The higher cost of metallocenes will probably hinder the use of these materials in this highly competitive and cost conscious market sector.

The use of LLDPE (Z-N and metallocene) in bag manufacture improves the following properties:

- Tensile strength
- Puncture resistance
- Work absorption
- Seal strength

22.2 Properties of heavy duty sack materials

A comparison of three LLDPE gas phase resins used for heavy duty films is shown in **Table 22.1**. The measurements were carried out on actual bags produced from the three resins. The data illustrates the superiority of the hexene resins compared to butene.

Table 22.1
Comparison of LLDPE heavy duty sack films

	Units	LLDPE (C_4)	LLDPE (C_6)	LLDPE (C_6)
Melt index	dg/min	0.8	1.0	0.8
Density	g/cm^3	0.921	0.918	0.926
Thickness	μm	190	150	150
Dart impact	g	750	1030	780
Dart impact	g/μm	3.95	6.87	5.2
Elmendorf tear MD/TD	g	1455/1370	2300/3145	2015/2100
Tensile yield MD/TD	MPa	10/10	11/11	13/13
Tensile strength MD/TD	MPa	28/29	37/38	35/38
Secant modulus MD/TD	MPa	200/200	220/230	290/300

Low slip level is necessary only on the outside of the bag to avoid sliding of the bags during transportation on the pallet. A COF of 0.6 has proven successful for many applications. For more special requirements, coextruded bags with EVA in the outside layer have been utilized.

High heat stability during filling with hot material is attainable with increased density LLDPE. In extreme cases of high filling temperatures, coextruded films with an HDPE core layer and LLDPE outside layers are a possible solution.

Most bags are printed. The resins usually contain pigment, thus screws with mixing sections should be used. To facilitate printing, the films are corona treated. A treatment level of 42 mN/m is recommended. A generator that can deliver at least 20 W/m^2/min. of net power at the electrodes (not input power) will be required.

22.3 Some options for heavy duty bag manufacture

Coextruded films are penetrating this market. These offer the advantage of placing a high COF layer on the outside of the bag for anti-slip characteristics, while avoiding blocking on the inside with an antiblock containing skin. For very demanding applications coextruded films of the following structures are used. The HDPE is added to increase stiffness and assist in hot filling.

Film #1	LDPE	LLDPE	LDPE
Film #2	LLDPE	LDPE	LLDPE
Film #3	LLDPE	HDPE	LLDPE

Coextruding with the optimum combination of polyethylenes will allow downgauging to 130-140 μm from 180 μm monolayer film and an effective saving of 22% raw material is possible. Film #2 is the more difficult to produce because of the need to use wide die gaps with LLDPE thus rendering the film less balanced as MD orientation is increased. Blending the outer layers with LDPE and the addition of PPA is probably necessary to optimize processing with this combination. PPA adds cost.

Other options for heavy duty sack manufacture include blends of LLDPE with HDPE as shown in **Table 22.2**. These are compared to a lean LD/LLDPE blend and a higher α-olefin (4-methylpentene-1) LLDPE for physical properties. A blow-up ratio of 1.75:1 was used in all experiments.

The polymers used in **Table 22.2** had the following basic characteristics:

LLDPE 4-methylpentene-1 (4MP1) (0.6 MI/0.920 g/cm^3)

LLDPE C_4 (0.9 MI/0.920 g/cm^3)

LDPE (0.3 MI/0.922 g/cm^3)

HDPE (0.7 MI/0.960 g/cm^3)

Table 22.2

LDPE and HDPE blends in heavy duty bag manufacture

Properties	LDPE/ LLDPE(C_4) 90/10 blend	LLDPE (4MP1)	LLDPE(4MP1)/ 7.5% HDPE	LLDPE(4MP1)/ 15% HDPE
Thickness (μm)	150	130	125	120
Impact strength (g)	600	1030	830	730
Tear strength MD (N)	7.5	18	14	10
Load @ yield MD/TD	16.5/16.5	15.1/15.9	15.5/16.2	16.2/17
Load @ break MD/TD	38/33	55/53	48/45	45/43
Relative stiffness	100	99	104	105

Table 22.3 show another set of data comparing the effect of blending two higher α-olefin LLDPEs in LDPE for heavy duty sack film. There is very little difference between hexene and octene comonomers.

Table 22.3

Hexene versus octene in heavy duty bag films

Property	Units	LLDPE (C_6) 40% LDPE	LLDPE (C_8) 40% LDPE
Melt index LLDPE	dg/min	0.8	1.0
Density LLDPE	g/cm^3	0.921	0.920
Thickness	μm	74	76
Dart impact	g	325	330
Elmendorf tear MD/TD	g	320/1160	375/1140
Tensile yield MD	MPa	11	10
Tensile strength MD	MPa	32	30
Elongation MD	%	590	570
Secant modulus MD	MPa	211	165

Adding 25% LLDPE to a fractional melt index LDPE can improve extruder output and will yield superior stress crack resistance. Adding 10% HDPE to LLDPE will improve bubble stability. Stiffness and creep resistance will also be enhanced. **Table 22.4** lists typical PE grades utilized in heavy duty sack applications. Many blending and coextrusion options enable property and cost optimization.

Table 22.4

Polymers for heavy duty bags

PE Type	Melt index (dg/min)	Density (g/cm^3)	Comonomer	Comment
LDPE	0.2	0.919	N/A	High impact LDPE
LDPE	0.3	0.922	N/A	Classic LDPE grade
LDPE	0.80	0.918	N/A	Medium duty
LDPE	0.2	0.924	N/A	Stiffness
LLDPE	0.5	0.918	C_4	Blending partner
LLDPE	0.8	0.926	C_6	Medium stiffness
LLDPE	0.9	0.920	C_4	Blending partner
LLDPE	1.0	0.918	C_4	Easier processing
LLDPE	1.0	0.926	C_8	Adds stiffness
LLDPE	1.0	0.920	C_8	Blending partner
mLLDPE	1.0	0.918	C_8	Blending partner
HDPE	0.28	0.945	C_4	Medium MWD
HDPE	0.03	0.95	HMW	Broad MWD
HDPE	MI_5 0.3	0.953	HMW	Broad MWD
HDPE	0.3	0.955	MI_5 1.4	Medium MWD
HDPE	0.7	0.960	MMW	Blending partner

Chapter 23

Surface protection and adhesive films

Two applications are reviewed in this market segment:

- Temporary protection film applied to shield flat surfaces
- Films to bond incompatible surfaces by the application of heat and pressure (thermal film)

23.1 Surface protection films

The film should typically protect sheet materials from scratches, dust and impact during transportation, shaping processes, cutting, installation etc. Typical materials that need such protection are:

- Metal (stainless steel, aluminium)
- Acrylic sheet (PMMA)
- Polycarbonate sheet (PC)
- Other plastic surfaces
- Wood or melamine laminated surfaces

The film must adhere to the surface of the substrate sufficiently, but must be able to be easily peeled off when ready for use or installation. In secondary processing such as cold stamping of stainless steel into kitchen sinks or other shaped products, the film must also resist the deformation (elongation) of the forming process as well as the oils used in metal forming techniques. The following properties are therefore critical:

- Tensile strength
- Puncture resistance
- No gels
- High elongation
- Controlled adhesion to the protected surface

Surface protection films have been traditionally produced from acrylic adhesive-coated blown PE films. Recently, for PMMA (Plexiglas) and polycarbonate

(Lexan) sheet there is a trend to use coextruded films with LDPE or LD/LLDPE blends and EVA or EAA to provide the required adherent surface.

For stainless steel sheet protection coextruded LD/LLDPE/ionomer films are also in development. In some applications HDPE films roller coated with an adhesive are also used. The film is applied for protection during handling, transportation and secondary processing.

23.2 Adhesive films

The films are heat laminated to bond two non-adherent materials such as plastics, textiles, metals, foamed rubber and papers without the use of liquid glue. Another use is as surface protection or to provide a heat-sealing layer to the substrate. The films are produced from polymers that have bonding properties under heat and pressure. Film thicknesses are in the range of 30-50 μm. The following resins are typically used:

- EVA (12-18% VA)
- EAA and ionomers
- Alkyl acrylates (EEA, EMA etc.)
- Terpolymers and MAH-grafted polyolefins

In addition to films this market segment includes nets and slit films that adhere to surfaces under heat and pressure. Key properties are:

- High adhesion strength to substrate
- High cohesive strength
- Good impact, puncture and tear strength
- Thermoforming in certain end uses
- Easy processing/laminating
- Resistance to dry cleaning and hot wash for garment applications
- Low and high temperature resistance for automotive applications

Acid-based copolymers, e.g. EAAs have very good adhesion to metals, polyurethane and rubber foam. Synthesizing a terpolymer with ethylene, acrylic acid and methyl acrylate will improve adhesion to non-metallic surfaces as well as retaining good bond strength to metals. Resins grafted with MAH groups have excellent adhesion to polyamide (PA) and polyester (PET) surfaces. The key markets are:

- Automotive
- Textiles
- Footwear
- Garments
- Carpet backing/underlay

For the protection of PUR (polyurethane) foam carpet underlay a 25 μm EAA/LDPE coextruded film is bonded by heat lamination to the bottom side of the underlay. This protects the flexible foam structure from damage during transportation and handling. In this application PE is only used to reduce the cost of the film.

Laminates for building, tunnel and container lining can be aluminium/aluminium or steel/steel bonded together with EAA, ionomers and MAH-grafted polyolefins. Resins with acidic functionality adhere best to metals.

Woven PP is bonded to PUR foam for use as wallpaper with E-EA-MAH terpolymer. High adhesion strength between PP/PUR should be obtained with PE-g-MAH or EMA-g-MAH.

Laminates are used in automotive applications as the inner lining of the boot/trunk, doors and rear panels. The thermoformed laminates can have the following structures:

- EPDM/adhesive/felt
- PUR foam/adhesive/felt
- Wood fibres/adhesive/woven PP or PET

In these applications, the adhesive layer is a 40-50 μm E-MA-MAH film. The comonomer content is approximately 11% b/w.

Other applications for adhesive films include bonding by heat lamination of paper/paper and paper/aluminium. EAA films have excellent adhesion to pulp and paper products and aluminium.

Chapter 24

Construction and building films

The building and construction industries account for 25% of polyethylene film consumption. One of the most important uses of polyethylene film in buildings is as a protective moisture barrier. Polyethylene film is used as a lightweight tarpaulin for covering building materials and equipment on site. Polyethylene sheet is also used for weatherproofing buildings under construction. The films used to cover equipment and building materials will be natural, i.e. not pigmented, to enable stored goods to be identified and when used as protection on scaffolding to maximize light transmission and visibility.

Other applications are in concrete pouring where the PE film is placed over the concrete to avoid loss of moisture and retain the heat of hydration. The low cost and functionality of LL/LDPE films cannot be matched by any other material in these applications.

Thickness usually ranges from 100-150 μm. Common film roll dimensions are 1 m x 30 m, but films as wide as 14 m are produced. Black films are utilized for membranes and water protection. The most common structures are:

- Monolayer LD/LLDPE blends
- Reinforced LD/LLDPE with woven oriented HDPE ribbons with fire retardant coating

Typical polyethylene resins used in building and construction applications are listed in **Table 24.1**. This is a very cost-sensitive market which uses much recycled materials to minimize costs.

Table 24.1
Polymer films for construction and civil engineering

PE Type	MI dg/min	Density g/cm^3	Comonomer	Comment
LDPE	0.2	0.922	-	Heavy duty film
LDPE	0.3	0.919	-	Heavy duty film
LDPE	0.3	0.934	Black	Opaque black film
LDPE	0.9	0.923	-	General purpose grade
EVA	0.5	0.926	4% VA	Toughness (low temp.)
EVA	1.5	0.926	2% VA	Toughness
LLDPE	0.5	0.918	butene	Used if cost lower than LDPE
LLDPE	1.0	0.926	octene	Usually blended
LLDPE	1.0	0.918	butene	Used if cost lower than LDPE
MDPE	0.6	0.942	butene	Toughness/stiffness
HDPE	0.055	0.950	Broad MWD	Stiffness/MVTR
HDPE	0.05	0.950	Broad MWD	Stiffness/MVTR

24.1 Roofing films and membranes

Two types of films/sheet are used in this market sector.

Vapour barrier roofing films are made from blends and coextrusions utilizing mainly LLDPE and HDPE. Films are usually micro-perforated to allow for breathing but preventing water penetration.

Membrane films are usually reinforced with a woven material and are used between the roof and ceiling to stop the penetration of moisture. Other applications include damp course membranes which prevent moisture rising up a building from the ground.

Roofing applications specify the following properties:

- Tear propagation resistance
- Toughness
- Moisture barrier
- Fire retardant requirements in specific cases

These films have to be produced to exacting specifications since the application is a permanent fixture in the construction. Reinforced black films are used with an oriented polyolefin scrim or woven material. The films have to be durable to resist handling and abuse on building sites during installation.

A wide range of materials are used for roofing applications, which include synthetic elastomers, PVC, and bitumen modified with EVA or SBS (styrene butadiene). The specifications can be demanding and crucially are very cost-sensitive. Modified bitumen is the most widely used material in roofing. These blends are also widely used in road surfacing.

Chapter 25

Hygiene and medical films

PE films are widely used in various health and personal care related applications from disposable gloves to hospital bed sheets and operating table covers, and personal hygiene products. The market segments discussed in this section are:

- Disposable gloves
- Diapers (nappies), incontinence pads and sanitary products
- Bags and hospital films
- Drainage bags, blood and IV (intravenous) fluid bags and general replacements for glass bottles

25.1 Disposable gloves

Gloves and related plastic protection products are used throughout the medical, cosmetics, food processing, institutional and pharmaceutical markets. Applications range from medical examination gloves to multi-purpose utility gloves.

Medical gloves are divided into two varieties, according to use. Surgical gloves are usually made from stretchable rubber latex. Medical inspection gloves are made from PE films that are hot stamped and welded around the edge of a flat hand shape. These are also used in point of sale applications, such as petrol (gas) stations. The films require the following key properties:

- Extensibility (for fit)
- Puncture resistance
- Seal strength
- Softness

EVA, EnBA and EMA are used for inspection gloves. The preferred materials are EnBA and EMA as they are less sticky than EVA and softer at equivalent comonomer contents. The films are usually produced by the blown film process. Film blowing is not easy. The film is around 25 μm in thickness and with 17-18% comonomer has very low modulus and is easily stretched. Great care must be taken particularly in the winding operation to avoid stretching of the film.

Relatively high levels of slip/antiblock and some filler are required to facilitate openability and machineability. To reduce costs EVA (7% VA) is blended with 10-20% LLDPE.

Gloves are hot stamped out of two layers of the film. Comparison of EnBA and EMA properties of blown film (both 17% b/w comonomer) indicate slight property differences as shown in **Table 25.1**.

The minor difference between the two copolymers is attributed to the different molecular size of the methyl acrylate (MA) and n-butyl acrylate (nBA). MA has a lower molecular weight than nBA. Therefore, for the same weight percent in the copolymer e.g. 17% in this case, there are more moles of MA than nBA copolymerized with ethylene.

The very low density mPEs are also possible candidates in this market, which requires soft stretchable films. The PE resins cannot be used in HF/RF (high or radio frequency) sealing and welding converting processes.

Table 25.1

EnBA and EMA for gloves

Property	EnBA (17%)	EMA (17%)
Yield strength	+	-
Tensile strength	-	+
Ultimate elongation	-	+
Secant modulus MD	same	same
Weld tensile strength	+	-
Glove elongation	-	+
Shrink/orientation	worse	better
Sealing temperature	higher	lower

25.2 Diapers, incontinence pads and sanitary products

The hygiene and personal care market has grown very rapidly in recent years. The market is continually expanding globally and providing opportunities for innovative products. The market is divided into three basic segments:

- Adult incontinence
- Feminine hygiene
- Baby diapers (nappies)

Hygiene films are generally constructed of three primary layers (**Figure 25.1**). The outer layer (backing) is a thin sheet of polyethylene film. Bound on this layer is an absorbent non-woven web of fluffy wood pulp, filled with super absorbents. The third inner layer (in contact with the skin) is generally a non-woven synthetic fibre such as polypropylene or polyester.

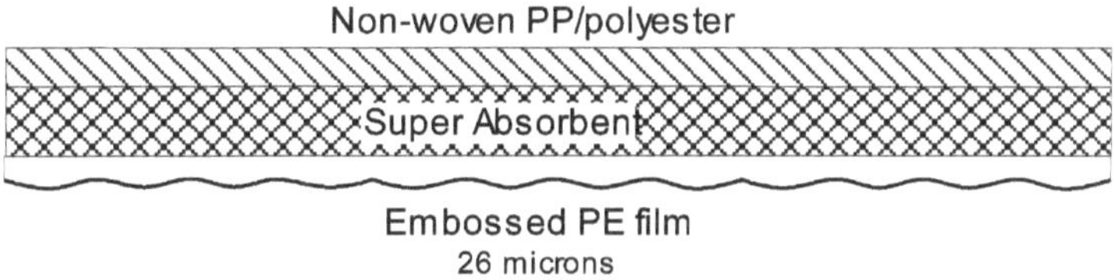

Figure 25.1

Basic structure for absorbent hygiene products

The back-sheet provides the base layer upon which the product is assembled and this is the ultimate water barrier. The latest films are multilayered, proprietary blends and micro-embossed to simulate a textile feel. Breathable films are made by compounding LLDPE with chalk and stretching the film to form micro-perforations. The following properties are particularly important:

- Impact strength to prevent burst at packing station
- Liquid barrier
- MD strength for high speed feed of the web
- TD strength required to reinforce the tape fastening area
- Soft feel and drape (controlled by embossing)
- High speed machineability requiring high mechanical demands
- Filler acceptance
- Low gel count
- Skin contact
- Low noise: important in adult applications

Processors are evaluating the potential for mLLDPE in various diaper and other hygiene applications. The main advantage so far identified for these materials is very low extractable levels at very low density. This makes it possible to develop very soft stretchable films with acceptable tensile properties that will comply with dermatological requirements and avoid any skin irritation or other health related problems.

25.3 Embossed films

Embossed films are used in a variety of applications including diapers, disposable garments, bed sheets, shower curtains and tablecloths.

Polyethylene film for diaper backing was originally produced on blown film equipment. In the early 1970s new resins were developed allowing producers to use the cast film process. The thickness range of diaper film is between 25 and 35 μm. The cast process (**Figure 25.2**) is preferred by many producers as it offers superior softness and gauge control. Another economic advantage of the cast process is that cooling and embossing are carried out in a single step, whereas with the tubular process the film is first cooled and is then reheated for the embossing step. The flat die is usually of the coat hanger type and feedblock systems are included for coextrusion. Lines equipped with 4 m wide dies are in commercial use.

The film is inspected for pinholes, slit in-line and fed to the winder. Corona treatment is used to improve adhesion to glues, closing tapes and inks. The latest state of the art machines are designed with web handling systems that allow automatic scrapless winding at speeds up to 500 mpm.

There are patents regarding the engraving of the embossing rolls. The designs are usually in a diamond configuration and can range from 1 mm to 0.2 mm in size. The design is critical since it will control the feel and softness of the film. In addition embossing patterns can affect tape adhesion, surface gloss and coefficient of friction (COF).

The primary disadvantage in using the cast process is that the film has less transverse directional tensile strength. This obstacle is partly offset by blending LLDPE with LDPE and by coextrusion. Higher α-olefin LLDPEs including metallocenes are effective blending partners. The very low density PEs are used as impact and softening modifiers. The actual blends and multilayered structures used for specific applications are proprietary and may be patented.

In applications for adult incontinence products, noise level is considered very important. These applications specify soft copolymers such as EVA, EMA and EnBA. Metallocene PEs are obvious candidates in these growing market segments. Embossed mPE (D: 0.905 g/cm^3) films were shown to have the following advantages over calendered plasticized PVC films:

- Higher tensile strength by some 20-25%
- Superior tear resistance
- Less film blocking

- Better MVTR worse for OTR
- The mPE offers opportunities for downgauging

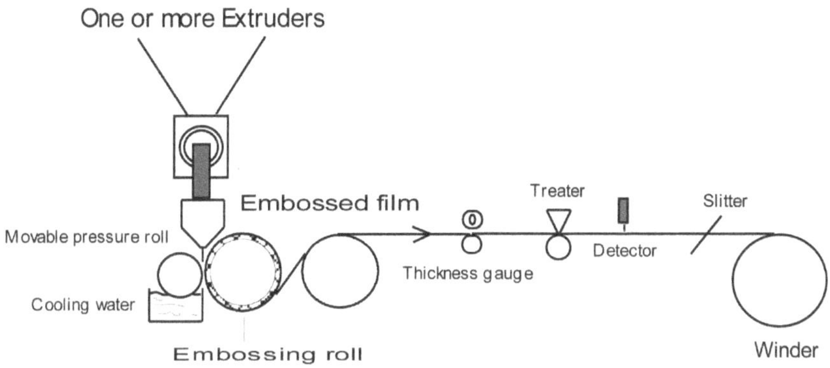

Figure 25.2

Embossed cast film layout

Extensive developmental work has been done to introduce coextruded films into this market. Coextrusion offers the possibility of using recycled resin in the core layer as well as greater flexibility in optimizing strength, feel, surface properties and cost. For instance by coextruding a core layer of a stiff material with skins from a softer polymer a film with an optimum balance of tensile strength and softness can be made. For many applications the COF has to be different from side to side, e.g. low for converting and non-slip in use. Surface characteristics are best controlled by the use of appropriate resins and fillers rather than migratory slip additives such as erucamide. The latter additives can migrate to all surfaces with time causing problems in adhesion and heat-sealability.

25.4 Hospital films

This sub-segment covers applications, such as temporary hospital bed drapes and particularly important covers for operating tables. The films are often coextruded usually white one side and green on the other. Disposable aprons are typically made from 20-25 μm HMW-HDPE film for hospital use which can be gamma radiation sterilized if required.

The films used as table covers or bed covering in this market sector need to be soft so that they can easily be draped on flat surfaces. Thickness will range from 40 to 80 μm. The films should satisfy the following requirement.

- Soft feel
- Good tear resistance
- Low noise
- High tensile strength
- Sterilizable (steam or irradiation)

25.5 Medical products

Plasticized PVC is one of the largest volume materials used in the manufacture of medical products such as blood, IV and drainage bags and catheter tubing. PVC films offer an impressive array of attributes, which are difficult for alternative polyolefin-based candidates to match or exceed at the same cost. The wide latitude of properties achievable by dry blending techniques using PVC powders with liquid plasticizers and stabilizers have been perfected over the years by processors leading to reluctance to switch to other materials. The extruders and dies are designed for PVC with its specific rheology and particularly die swell, which differ from polyethylenes.

The key features of plasticized PVC films for medical products (catheters and drainage bags) are:

- Low cost
- Elasticity
- Kink resistance
- Transparency
- Strength and durability
- RF welding (27.12 MHz)
- Solvent bonding
- Sterilized by autoclave, ethylene oxide (EtO) and gamma rays
- Wide service temperature
- Chemical resistance
- High melt strength: easy to extrude

To the above must be added a long and successful track record in medical applications. It is important to remember that steam sterilization (autoclaving) cannot be used with low density polyethylenes and neither can RF welding be used to seal through thick sheet.

Applications such as ostomy bags and the fluid drainage market are rapidly increasing with ageing populations. Some key requirements are listed below:

- Flexibility
- Low noise
- Seal strength
- Odour barrier
- No skin irritation

An OTR barrier of 20 cm^3/m^2/day is sufficient for this application. Lightly plasticized PVDC provides sufficient barrier and flexibility (low noise) and can be coextruded with EVA (16-18% VA) without a tie-resin. The EVA can be modified by the addition of EPDM elastomer plus filler to add flexibility and reduce blocking. If EVA causes skin irritation, a very low density mPE may be considered as an alternative candidate. Metallocenes have lower levels of extractables and form stronger films. However, adhesion to PVDC is poor compared to EVA.

Low density PE (mPE) can be used in five layered coextrusions with a barrier resin, i.e. PVDC or EVOH although some sources have indicated that EVOH is too stiff for this application. A stiff film will generate embarrassing noise when worn on the body. The use of 5-layer EVOH structures will probably prove too costly.

The results in **Table 25.2** and **25.3** comparing mPE to soft PVC have been extracted from data published by Dow Chemical Co. The mPEs tested were based on resins with octene contents of 12-20%, i.e. density 0.88-0.90 g/cm^3. The films were embossed and made by a cast film (monolayer) process. The PVC films were commercial medical grade materials.

Table 25.2 shows some typical properties of medical films made from both materials. The barrier properties of the vinyl films are a function of their plasticizer content and presence of other additives. The mPEs in these comparisons have OTR values that are approx. 4 times higher than the PVC films and MVTR values that are approx. 5 times lower. Depending upon the application a high OTR may be beneficial for breathability or problematical due to odour transmission.

Table 25.3 shows some data comparing medical collection bags made from mPE and PVC films. The tests are based on standard methodology to evaluate liquid collection bags. The bags were fabricated from two sheets of film welded together by RF. Surface area was 432 cm^2. To overcome the low dielectric loss of mPE which normally precludes RF welding, the films were welded with a reusable "catalyst" film with the RF equipment stabilized at 27.12 MHz.

Table 25.2

Properties of mPE and PVC films

Property	Unit	mPE 1	mPE 2	PVC 1	PVC 2
Thickness	mm	0.15	0.15	0.18	0.25
Density	g/cm^3	0.905	0.895	1.253	1.260
Tensile yield MD/TD	MPa	5.0/5.0	4.0/4.1	8.8/7.1	6.0/5.4
Elong. @ break MD/TD	%	700/705	625/640	185/330	270/350
1% Sec. modulus MD/TD	MPa	62.8/61.2	43.4/43.5	51.8/49.5	24.4/36.7
Elmendorf tear MD/TD $x10^3$	g/mm	11/13.7	5.3/6.7	12.0/12.6	4.2/4.1
Puncture resistance	J/m^2	9.2	14.3	7.17	7.0
OTR	$cm^3/m^2/day$	2635	3490	620	-
MVTR	$g/m^2/day$	6.5	8.6	-	-

In **Table 25.3**, the mPE films compare very favourably with their vinyl counterparts particularly in the impact drop test which is influenced by the high elongation to break inherent in linear polyethylenes. Interestingly some of the data was shown not to be directly related to film thickness.

The burst strength data shown in **Figure 25.3** was determined by filling the bags with air at a controlled pressure and flow rate and measuring the time to failure. Filling rate was 5.36 L/min at a line pressure of 0.083 MPa until the bags burst. **Figure 25.3** shows the superior resistance of the mPEs in the air burst test and in particular that of the very low density mPE 2 film.

Table 25.3

Properties of medical device films

Property	Unit	mPE 1	mPE 2	PVC 1	PVC 2
Thickness	mm	0.15-0.25	0.15-0.25	0.18-0.23	0.25
Density	g/cm^3	0.905	0.895	1.25	1.26
Air burst test time	s	170-184	288-269	104-72	105-73
Resistance to bursting	N	1400->2300	>2300->2300	>2300	>2300
Resistance to impact	m	>7.6	>7.6	2.5-2.7	2.7-3.1

The disadvantage of very low density PE is a melting point <110°C, which eliminates heat sterilization as an option. However, these polyolefins can be sterilized by gamma irradiation without yellowing, or by ethylene oxide. Also, solvent bonding commonly used for joining PVC components will not work well with mPE films.

Applications such as collection and drainage bags and inflatable devices have been suggested for PE films. As a general conclusion, to date very few inroads has been made by polyolefins to replace vinyl films in medical products such as blood and IV bags and catheter tubing. However, there is still a drive to eliminate chlorinated materials both in use and from waste streams.

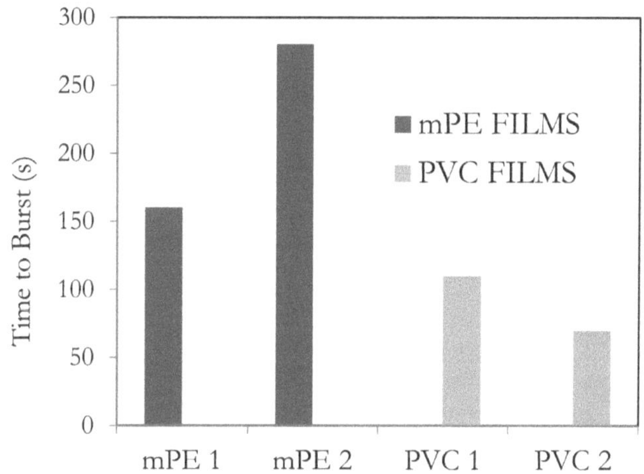

Figure 25.3

Comparative burst resistance of mPE and PVC films

Chapter 26

Other PE film applications

In this final chapter several demanding applications for polyolefin films are discussed, such as skin packaging, shrink sleeves, labels and bubble films.

26.1 Skin packaging films

Protection of the product is generally assumed as being paramount in all packaging techniques. However the role of packaging in consumer marketing is also frequently to attract, entice, convince and make the sale. Skin packaging, like its near relation shrink packaging will perform this function.

In skin packaging the article is placed on a porous support and fed to the packaging machine. Film is drawn from a reel and clamped in a frame set above the loaded support. A cutter severs the film from the reel and a heater moves above the film. After a set time, the heater retracts and a frame descends, draping the hot film over the article. A vacuum is applied beneath the support and the film is drawn around the article into intimate contact and simultaneously sealed to the support.

The films described in this chapter are limited to extruded mono or multilayer films without gas barrier. These are used mainly in the packaging of consumer goods, e.g. hardware, toys, electrical and auto parts and short shelf-life food. The support of the package can be cardboard or other porous sheet.

PE films used in skin packaging should generally meet the following criteria:

- Seal strength/adhesion to the support
- Puncture resistance
- Finger indent recovery
- Clarity
- Gloss and sparkle
- High melt strength
- No wrinkles on the pack after the shrink process

Where indent recovery (snapback) is required none of the polyolefins will match plasticized PVC films.

Blown monolayer and coextruded film processes are normally used. The films are produced with a high blow-up ratio. The materials listed below may be used depending on the application and support material:

- Ionomers (Zn^{++} or Na^{+} cations)
- EAA
- EVA
- E-alkyl acrylate-MAH terpolymers
- Metallocene PE

In general, because of their superior heat-seal properties, higher puncture strength and melt strength, ionomers and EAA are preferred. Neither of these materials require pre-treatment of the cardboard to ensure good adhesion. Some blown film converters have been developing low cost LDPE/EVA (12-15% VA) coextruded films for untreated cardboard supports. However, EVA does not exhibit the same toughness and puncture resistance as ionomer, and also with more than 9 wt.% VA resins the converter may encounter blocking and machineability (low stiffness/stretch) problems.

These resins are used either as monofilms or coextrusions are shown in **Table 26.1**. The ionomers are generally considered to be the most effective but they are expensive for many applications.

Table 26.1

Polymers for skin packaging

PE Type	Melt index (dg/min)	Density (g/cm³)	Comonomer	Comment
EVA	10	0.926	5% VA	Toughness/clarity
EVA	4.0	0.929	9% VA	Used as tie-layer with PA/PE
EVA	14	-	14% VA	Coextrusion skin adhesive layer
EVA	6.0	-	14% VA	Coextrusion skin adhesive layer
EAA	2.5	0.932	6.5% AA	Good adhesion to cardboard
Terpolymer	1.0		E-MA/MAH	Coextrusion
Ionomer	1.3	0.94	Na^{+}	Puncture resistance
Ionomer	1.9	0.941	Zn^{++}	Coextrusion
VLDPE	1.0	0.905	C_8	Coextrusion: mPE preferred

26.2 Labels and shrink sleeves

Labels are used in packaging, mailing, automatic dispensing, etc. Various materials such as paper, PEs, BOPP, PVC, PS, etc. are used in this application. In addition to paper, LDPE and MDPE, and for shrink-sleeves LDPE are the major films used for label manufacturing. The pressure-sensitive adhesive is normally applied from hot melt formulations. These are based on blends of wax, tackifier resin and a polymer such as EVA. Critical film properties are:

- Good optical properties (gloss, clarity)
- Excellent print surface
- Easy dispensing and good machineability (high speed application)
- Compatibility with adhesives
- Shrink and/or stretch properties
- Recyclability without separation from the packaging material

The majority of PE-based labels are mono-films, pressure-sensitive or shrink-sleeve types as shown in **Table 26.2**.

Table 26.2

Films for labels

Label Types	Description
Pressure-sensitive	LDPE or MDPE films, 60 to 90 μm thick, corona-treated, printed and coated with a curable or non-curable acrylic-based pressure-sensitive adhesive
Shrink-sleeve	Mono-oriented LDPE or MDPE films, 100 to 120 μm thick, corona-treated and printed

In comparison to pressure-sensitive labels, the advantages of shrink-sleeve products are lower cost, no glue and fewer recycling problems.

Other PE-based labels are:

- Spun bonded fibre HDPE sheet (Tyvak®)
- Microcellular PE sheet

The main criticisms of plastic films are that they are difficult to print and are more expensive than paper. However, with the right ink formulation and printing conditions, inks are able to dry and adhere in a reasonable time.

The most common flat label film is made from HDPE. Film may be single or multilayer coextruded (HDPE/PP/HDPE) and there are various manufacturing techniques to achieve machine/cross direction equilibrium. Films may be filled and/or coated with pigment, kaolin, calcium carbonate and titanium dioxide, and may be coloured, all to give paper-like characteristics of appearance, feel, and converting. Normal thickness ranges from 60 to 100 μm. Labels thinner than 60 μm are difficult to peel off the silicon-coated carrier paper.

HDPE offers the following physical characteristics that are required by the printer:

- Flexibility: to enable the label to be easily shaped round small diameters
- Rigidity: for ease of handling and label dispensing
- Dimensional stability: for flatness and moisture resistance ensures that there are no dimensional changes with humidity
- Strength: aids waste stripping
- Opacity: the opacity can often be increased if necessary
- Printability: both for inks keying and speed of drying

Excellent chemical and oil resistance of HDPE films protect the label from damage in unfriendly environments. HDPE labels are widely used in a number of applications such as:

- Labels for chemical drums
- Horticultural labels, tags and ties
- Where variable information is imparted, e.g. airline luggage tags

In-mould labelling for injection moulded lids and pots for margarine and ice cream is well established. In this process the labels are placed in the mould and are subjected to the same process conditions of heat and pressure as the moulded container. It is therefore preferable for processing and recycling to use materials similar to those of the container unless recycling considerations demand separation of the printed label from the unprinted container. Some other plastic films used for labels are listed below:

- Polyethylene (LDPE/HDPE): clear, pigmented, matt/gloss surface treated and top-coated
- Polypropylene: clear, pigmented, matt/gloss surface treated and top-coated
- Polystyrene: clear, pigmented, matt and gloss surface treated
- Plasticized PVC: matt and gloss
- Rigid PVC: matt and gloss
- Polyester: matt and gloss, surface treated and top-coated

26.3 Bubble film

The manufacture of bubble film, developed by Sealed Air Corporation, is an ingenious process to produce a flexible film based on the entrapment of air to provide a cushioning effect to absorb energy.

Two PE films extruded from flat dies are conveyed into a roll stack. The bottom roll (suction roll) is drilled with holes ranging from 6.0-25.4 mm. A vacuum is applied and the hot film is sucked into the cavities. The second film layer is then drawn over the cavities and heat-sealed to entrap the air and form the bubbles.

The two film layers can be applied from one extruder with the melt stream split at a T junction to feed the two dies, or from two separate extruders if different polymers or blends are specified. Some machines can run at line speeds in excess of 1000 mpm.

The films are typically LLDPE blends and coextrusions. Recycled materials are also used. For some applications a barrier layer is added to reduce the air permeability of the film. These barrier layers have varied from PVDC (Saran) coating to coextrusions with PA and EVOH.

Bubble films are widely used to protect a variety of fragile items, such as glass, ceramics, furniture and delicate electronic equipment. For some applications antistatic films are recommended. Their flexibility enables them to be wrapped round articles or used flat in containers or envelopes.

There are a number of variations on the basic structure of the bubble. Some examples are shown in **Figure 26.1**. Example A is the basic structure with the PE films sealed together entrapping the air in the cavity. In B other films and composites can be laminated to the basic PE structure to enhance performance for specialized applications. These composites can include non-woven fabrics,

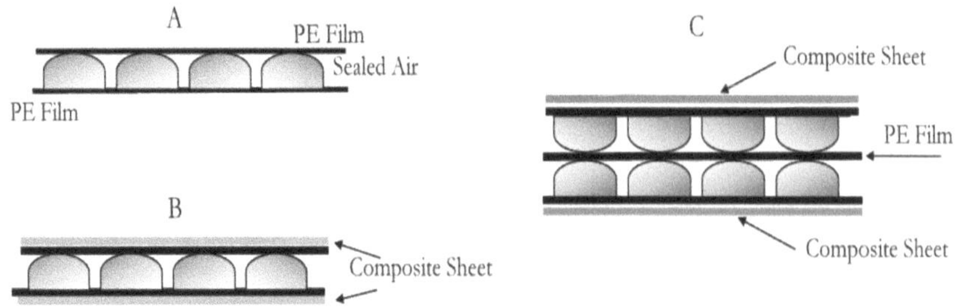

Figure 26.1

Various structures of bubble films

metallized films and aluminium foil, reinforced papers, conductive films and other options.

Example C shows a double bubble structure laminated with PE film to increase the cushioning effect. These can also be laminated to various composites.

Recommended further reading

Dooley, J. & Hilton, B.T., Layer Rearrangement in Coextrusion, Plastics Engineering, **1994**, 50, 2, p.25

van Krevelen, D.W. & te Nijenhuis, K., Properties of Polymers, 4th Edition, **2009**, Elsevier Science, 978 0080548197, Ch. 7

Detzel, A., Wellenreuther, F. & Kunze, S., LCA of waste bags, **2009**, Institut für Energie- und Umweltforschung (IFEU) Heidelberg (http://www.plasteurope.com/news/WASTE_BAGS_t214249)

Baird, W.G., Lindstrom, C.A., Besse, A.L. & d'Entremont, D.J., Method of producing film having improved shrink energy, US Patent 3,022,543, **1962**

Benning, C.J., Gregorian, R. & Werber, F.X., Process for forming cross-linked polyethylene film, US Patent 3,201,503, **1965**

Zisman, W.A., Contact Angle, Wettability and Adhesion, ACS Advances in Chemistry series, **1964**, 43, p.1-51

Tracton, A.A., ed., Coatings Technology Handbook, 3rd Edition, **2005**, CRC Press, 9781574446494

Kollbach, G. & Sauer, R., Polyurethane Lamination Adhesive. US Patent 20100010156, **2010**

Dooley J., Determining the Processability of Coextruded Structures, TAPPI Polymers, Laminations and Coatings Conference, St. Louis, **2007**, TAPPI Press, p.480

Schrenk, W.J., Bradley, N.L., Alfrey Jr., T. & Maack, H., Interfacial Flow Instability in Coextrusion, Polymer Engineering and Science, **1978**, 18, p.620-623

Helmy H., Recent Developments in Flat Film and Sheet Dies, Plastic Film and Sheeting, **1987**, 4, p.193

Rauwendaal, C., Polymer Extrusion, 5th edition, **2014**, Hanser Verlag, 9781569905166

Fischer, P. & Wortberg, J., New Spiral Die Designs, Innovations in Extrusion Conference, SKZ, **1998**

Dooley, J., Robacki, J., Jenkins, S., Lee, P.C. & Wrisley R., Producing Microlayer Blown Film Structures Using Layer Multiplication and Unique Die Technology, Antec 2011, SPE, **2011**

Shelat , K.J., Dutta N.K. & Choudhury, N.R., Interfacial Interaction and Morphology of EVOH and Ionomer Blends Using Scanning Thermal Microscopy and its Correlation with Barrier Characteristics, Langmuir, **2008**, 24 p.5464

Fetell, A.I., Nylon-Containing Ionomer Improves the Performance of EVOH Copolymers in Multilayer Structures, TAPPI Polymers, Laminations and Coatings Conference, **1997**, TAPPI Press, p.503

Sacharow, S. & Griffin, R.C., Principles of Food Packaging 2nd Edition, **1980**, AVI Publishing, 9780870553479

Graeff, R.W., Getting Optimum Results from PET Drying, Mod. Plastics Intl, **1992**, May, p.99

Pechinski, L.J., Changing Demands Spur Adhesive Developments, Paper Film and Foil Converter, **1991**

Bezigian, T., The Effect of Corona Discharge on Polymer Films, TAPPI Journal, **1992**, 75, March, p.139-141

Cogswell, F.N., Polymer Melt Rheology During Elongational Flow, ACS Symposium, Philadelphia, **1975**

Brodie, V., Taste Testing as it Relates to Packaging Resins, TAPPI Journal, Dec. **1988**, p.160-162

Abad, M.J., et al, Use of a sodium ionomer as compatibilizer in polypropylene-barrier EVOH copolymer blends, Journal of Applied Polymer Science, **2004**, 94, p.1763-1770

Chan, C.M., Polymer Surface Modification and Characterization, **1993**, Hanser Verlag, 9783446158702

d'Agostino, R., et al., Plasma Treatment of PET for Improving Al-Adhesion, 41st Annual Technical Conference Proceedings, SVC, **1998**

Arefi-Khonsari, F., Tatoulian, M., Kurdi, J. & Amouroux, J., Study of Plasma Treated Polymers and the Stability of the Surface Properties, Joint International Meeting ECS and ISE, Paris, **1997**

Blong, T.J., Klein, D.F., Pocius, A.V. & Strobel, M., The Influence of Polymer Processing Additives on the Surface, Mechanical, and Optical Properties of LLDPE Blown Film, Journal of Plastic Film and Sheeting, **1994**, 10, 2, p.103-115

Agalloco, J., Akers, J. & Madsen, R., Aseptic Processing: A Review of Current Industry Practice, Pharmaceutical Technology, **2004**, October, p.126

Glossary of abbreviations

4MP1	4-Methylpentene-1
AA	Acrylic acid
APET	Amorphous polyethylene terephthalate
APP	Atactic polypropylene
BA	Butyl acrylate
BIB	Bag-in-box
BOPA	Biaxially oriented polyamide
BOPP	Biaxially oriented PP
BUR	Blow-up ratio
CD	Composition distribution
COF	Coefficient of friction
EAA	Ethylene-acrylic acid copolymer
EBA	Ethylene-butyl acrylate copolymer
EBA-g-MAH	Maleic anhydride grafted EBA
EEA	Ethylene-ethyl acrylate copolymer
EMA	Ethylene-methyl acrylate copolymer
EMAA	Ethylene-methacrylic acid copolymer
EnBA	Ethylene-n-butyl acrylate copolymer
EPDM	Ethylene-propylene-diene rubber
EVA	Ethylene-vinyl acetate copolymer
EVOH	Ethylene-vinyl alcohol copolymer
FEM	Finite element method
FFS	Form, fill and seal
GMO	Glycerol mono-oleate
GTR	Gas transmission rate
HAO	Higher alpha olefins
HDPE	High density polyethylene
HF	High frequency
HFFS	Horizontal form, fill and seal
HMW	High molecular weight
IBC	Internal bubble cooling
ID	Inside diameter

IQF	Individually quick frozen
LCB	Long chain branching
LD ratio	Length:diameter
LDPE	Low density polyethylene
LLDPE	Linear low density polyethylene
LLDPE-g-MAH	Maleic anhydride grafted LLDPE
MA	Methyl acrylate
MAA	Methacrylic acid
MAH	Maleic anhydride
MAP	Modified atmosphere packaging
MD	Machine direction
MDI	Diphenylmethane diisocyanate
MDPE	Medium density polyethylene
MEK	Methyl ethyl ketone
MI	Melt index
mLLDPE	Metallocene catalysed linear low density polyethylene
MMW	Medium molecular weight
MVTR	Water vapour (moisture) transmission rate
MW	Molecular weight
MWD	Molecular weight distribution
nBA	n-Butyl acrylate
NPE	Nonylphenol ethoxylate
OD	Outside diameter
OPET	Oriented polyester
OTR	Gas (oxygen) transmission rate
PA	Polyamide
PB	Polybutene
PE	Polyethylene
PE-g-MAH	Maleic anhydride grafted PE
PET	Polyethylene terephthalate (polyester)
PIB	Polyisobutylene
PID	Proportional-integral-derivative (control)
POE	Polyolefin elastomer
PP	Polypropylene
PPA	Polymeric processing aid

PPRC	Polypropylene random copolymers
PVC	Polyvinyl chloride
PVDC	Polyvinylidene chloride
RF	Radio frequency
SBS	Styrene-butadiene-styrene block copolymer
SCB	Side chain branching
SEBS	Styrene-ethylene/butylene-styrene block copolymer
SiO_x	Silicon oxide
SIT	Seal initiation temperature
TD	Transverse direction
TDI	Toluene diisocyanate
TPE	Thermoplastic elastomer
UV	Ultraviolet
VA	Vinyl acetate
VFFS	Vertical form, fill and seal
Z-N	Zeigler-Natta

Index

M

N

O

T

MIX
Papier | Fördert gute Waldnutzung
FSC® C083411

Kolibri 360 GmbH
Ferdinand-Jühlke-Straße 7
99095 Erfurt, Deutschland
produktsicherheit@kolibri360.de